SpringerBriefs in Electrical and Computer Engineering

For further volumes:
http://www.springer.com/series/10059

Jacob Benesty · Jingdong Chen
Emanuël A. P. Habets

Speech Enhancement in the STFT Domain

Springer

Prof. Dr. Jacob Benesty
INRS-EMT
University of Quebec
de la Gauchetiere Ouest 800
Montreal, QC H5A 1K6
Canada
e-mail: benesty@emt.inrs.ca

Emanuël A. P. Habets
International Audio Laboratories Erlangen
Univeristy of Erlangen-Nuremberg
Am Wolfsmantel 33
91058 Erlangen
Germany
e-mail: e.habets@ieee.org

Jingdong Chen
Northwestern Polytechnical University
Youyi West Road 127
710072 Xi'an
People's Republic of China
e-mail: jingdongchen@ieee.org

ISSN 2191-8112
e-ISSN 2191-8120
ISBN 978-3-642-23249-7
e-ISBN 978-3-642-23250-3
DOI 10.1007/978-3-642-23250-3
Springer Heidelberg Dordrecht London New York

Library of Congress Control Number: 2011937831

Cover design: eStudio Calamar, Berlin/Figueres

Printed on acid-free paper

Springer is part of Springer Science+Business Media (www.springer.com)

Contents

Chapter 1
Introduction

Noise is everywhere. In all applications that are related to voice and speech, from sound recording, cellular phones, hands-free communication, teleconferencing, hearing aids, to human-machine interfaces, a speech signal of interest captured by microphone sensors is always mixed with noise. Depending on its level, the noise can significantly contaminate the statistical characteristics and spectrum of the desired speech signal. To illustrate this, Fig. 1.1 shows a clean speech signal, the same signal recorded in a noisy conference room, and their spectrograms. It is seen that the presence of noise does not only add new frequency components, but masks a great portion of the time-varying spectra of the desired speech as well. Furthermore, the dynamic properties of the desired speech spectra near phonetic boundaries are smeared and the intermittent nature of speech becomes less distinct. These changes can greatly affect our perception of the desired speech when listening to the noisy one. On the positive side, we can still perceive the useful information embedded in the desired speech signal if the noise level is not too high; but it would take more attention and may easily lead to listening fatigue. On the negative side, it may become impossible to comprehend the desired speech if noise is too loud. Therefore, it is highly desirable, and sometimes even indispensable, to "clean up" the noisy signals captured by a speech communication and processing system before they are transmitted or played out. This cleaning process, which is often referred to as either speech enhancement or noise reduction, has become an important research and engineering problem for many decades.

In the literature, research in speech enhancement has been divided into two categories: single-channel and multichannel. Early studies were mainly focused on the single-channel scenario as most communication devices at that time were equipped with only one microphone. The basic problem with this case is to estimate the desired speech signal from its microphone observation, which is a mixture of the clean speech and noise. The estimation is typically accomplished by passing the noisy signal through a filter. Since both the clean speech and noise are filtered at the same time, the most critical, yet most challenging issue of speech enhancement becomes one of how to find a proper optimal filter that can significantly mitigate the noise effect

J. Benesty et al., *Speech Enhancement in the STFT Domain*,
SpringerBriefs in Electrical and Computer Engineering,
DOI: 10.1007/978-3-642-23250-3_1,

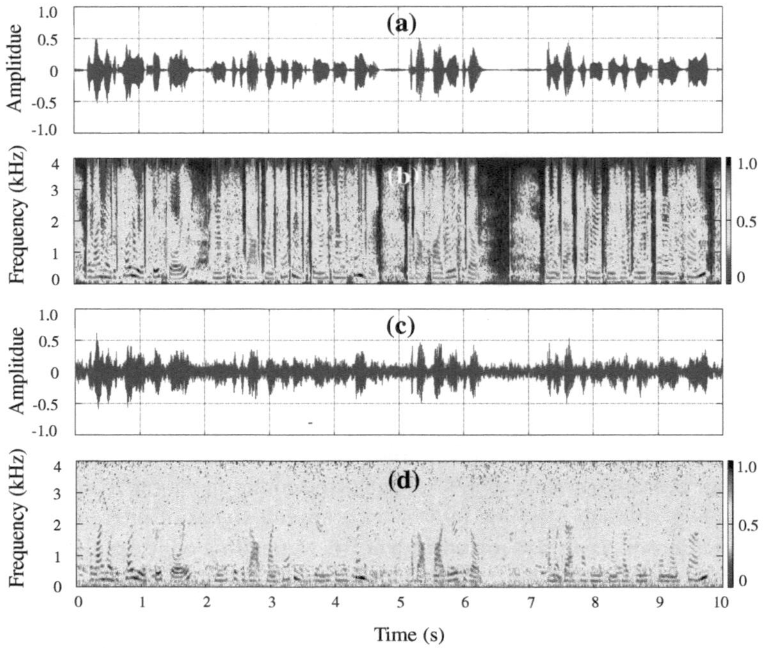

Fig. 1.1 Illustration of the noise effect. **a** A clean speech signal sampled at 8 kHz, **b** the clean speech spectrogram, **c** the noisy speech recorded in a conference room, and **d** the noisy speech spectrogram

while maintaining the filtered speech signal spectrally and perceptually close to its original form. The optimal filter can be designed in the time domain but on a short-time basis due to the fact that speech is highly nonstationary. The basic paradigm is to divide the observed noisy signal into short-time frames, typically in the order of a few to a few tens of milliseconds. For each frame in which signals can be treated as stationary, the statistics of the noisy and noise signals are estimated and an optimal filter is then constructed following the seminal work of Norbert Wiener [1]. Finally, an estimate of the clean speech is obtained by passing the frame of the noisy speech through the constructed filter. This approach, though theoretically sound and viable, is computationally expensive to implement as it often involves the computation of matrix inversion.

One way to circumvent the complexity issue is through the use of the fast Fourier transform (FFT), by converting the original problem into the short-time Fourier transform (STFT) domain. Now the enhancement process is accomplished in four basic steps. First, the noisy speech is divided into short-time frames as performed by the time-domain approaches. Second, each frame is transformed into the frequency domain via the FFT. This step is often called the analysis part of the process. Third, in each frequency band, a noise reduction filter is designed and applied to the complex

STFT coefficients to obtain an estimate of the clean speech spectrum. Finally, the time-domain enhanced speech is synthesized from the estimated clean speech spectrum using the inverse FFT (IFFT). Besides being computationally efficient, this structure of speech enhancement handles different frequencies independently, which gives an appealing flexibility in exploiting the noise statistics and using our knowledge on speech perception to optimize the enhancement performance. As a result, most efforts in speech enhancement in the past have been devoted to this framework. A brief review of such efforts is provided in the next section.

1.1 Single-Channel Speech Enhancement in the STFT Domain: A Brief Review

Research on single-channel speech enhancement in the frequency domain has been conducted for more than four decades during which many basic concepts and fundamental techniques were developed. This section briefly reviews some research highlights and milestones from our perspective.

The earliest speech enhancement system in the frequency domain was developed by Schroeder at Bell Laboratories in the 1960s [2, 3]. The schematic diagram of his system is shown in Fig. 1.2. Note that this diagram is slightly modified from its original form in [2] for ease of exposition. As seen, the input noisy signal $y(t)$ (where t denotes continuous time), which is supposed to be a superposition of the clean speech $x(t)$ and noise $v(t)$, is divided into K subbands. For each subband, a rectifier and a lowpass filter are applied in tandem to estimate the noisy speech envelope. The noise level in the corresponding subband is then estimated using an analog circuit with resistors, capacitors, and diodes, and the noise estimate is subsequently subtracted from the noisy speech envelope, resulting in an estimate of the clean speech envelope for the subband. A second rectification process is applied to force the negative results due to the subtraction to zero. The rectified clean speech envelope estimate, which is served as a gain filter, is then multiplied with the unmodified subband signal. Finally, the fullband signal $z(t)$ is constructed from all the subband outputs, where $z(t)$ is basically the estimate of $x(t)$. This work can be viewed as the first prototype of the well-known spectral magnitude subtraction approach, but with all analog implementation. The first endeavor that explicitly formulated the speech enhancement problem in the STFT domain was perhaps the famous spectral subtraction method developed by Boll in his informative paper published in 1979 [4]. Shortly after Boll's work, McAulay and Malpass cast the spectral subtraction approach in a framework of statistical spectral estimation, and presented a broad class of estimators including magnitude and power subtraction, Wiener filtering, maximum likelihood (ML) envelope estimator, etc. [5]. They were also the first to make connections between spectral subtraction and the Wiener filter. Almost at the same time, Lim and Oppenheim, in their landmark work [6], presented one of the first comprehensive treatments of methods of noise reduction and speech enhancement. The spectrum subtraction methods were discussed, also within an estimation

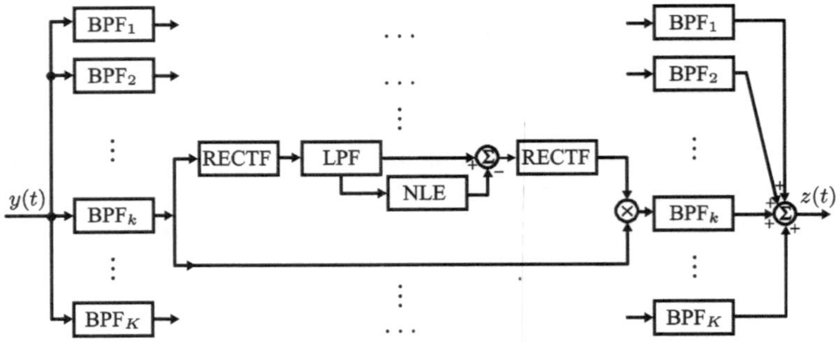

Fig. 1.2 Schematic diagram of the Schroeder's noise reduction system where BFP, RECTF, LFP, and NLE denote, respectively, bandpass filter, rectifier, lowpass filter, and noise level estimator

framework, and were compared to other techniques of speech enhancement. It was demonstrated that speech enhancement is not only useful in improving the quality of noise-corrupted speech, but also useful in increasing both quality and intelligibility of linear prediction coding (LPC)-based parametric speech coding systems. Also, Sondhi, Schmidt, and Rabiner published some results from a series of implementation studies that roots to Schroeder's work of the 1960s [7].

Though theoretically straightforward and practically effective, the spectral subtraction approach in general does not have optimality properties associated with it. In the 1980s, a great deal of efforts were devoted to finding optimal noise reduction filters in the framework of statistical estimation theory and several fundamental ideas surfaced and were published. The most notable work was the minimum mean-square error (MMSE) estimator for spectral amplitude (MMSE-SA) introduced by Ephraim and Malah [8]. It did not only give an optimal spectral amplitude estimator from the MMSE sense, but also demonstrated, by theory, that the optimal estimate of the clean speech spectral phase is the spectral phase of the noisy speech. Therefore, the basic problem of noise reduction becomes one of estimating only the clean speech spectral amplitude (this result was used previously, but mainly based on experimental observations). Following this work, many statistical spectral estimators were developed, including the MMSE log-spectral amplitude (MMSE-LSA) estimator [9], the ML spectral amplitude estimator [5], the ML spectral power estimator, the maximum a posteriori (MAP) spectral amplitude estimator [10], etc. Today, there are still tremendous efforts by contributors to find better spectral amplitude estimators.

To derive the noise reduction filters, the aforementioned MMSE, ML, and MAP estimators assume explicit knowledge of the marginal and joint probability distributions of the clean speech and noise spectra, so that the conditional expected value of the clean speech spectrum, given the noisy speech spectrum, can be evaluated. However, the assumed distributions may not accurately reflect the behavior of the real signals in practice. One way to circumvent this issue is to collect some speech and noise samples and learn the distributions from the collected data. This has led to the development of the hidden Markov model (HMM) based speech enhancement technique. HMM is a statistical model that uses a finite number of states and the associated state transitions to jointly model the temporal and spectral variation of signals [11]. It has long been used for speech modeling with applications to speech recognition [12–14]. The HMM was introduced to deal with the STFT-domain speech enhancement problem in the late 1980s [15–17]. This method estimates the optimal noise reduction filters in two steps. In the first step, which is often called a training process, the probability distributions of the clean speech and the noise process are estimated from given training sequences. The estimated distributions are then applied in the second step to construct speech enhancement filters. Similar to the traditional STFT-domain techniques, the HMM method also applies a gain to the noisy speech spectrum to reduce noise and many different gains can be formed [17, 18]. Besides not requiring explicit knowledge of the speech and noise distributions, the HMM technique has another advantage of being able to tolerate some nonstationarity in noise, depending on the amount of the training noise data and the number of states and mixtures used in the noise HMM. But distortion will arise when the characteristics of the noise are not represented in the training noise data.

All the aforementioned approaches, whether they are formed purely using empirical knowledge or derived from the framework of estimation theory, make a common assumption that the STFT coefficients at different frames and subbands are independent. Under this assumption, the optimal filter for any given frame and subband is basically a magnitude gain, which does not improve the frame-wise subband SNR (note that the long-time fullband SNR can be improved because the frequency-dependent and time-varying gain tends to lifter the subbands and frames that are less noisy and weigh down on those that are more noisy). A legitimate question one would ask: are the STFT coefficients from different frames independent? This will be answered in the next section and it will be shown that there exists strong correlation between STFT coefficients from neighboring frames.

1.2 Interframe Correlation

One major issue with the STFT-domain speech enhancement approach is the aliasing problem caused by circular convolution. To solve this problem, we need to use either the overlap add or overlap save techniques. (Note, however, that even with overlap add/save procedure, aliasing cannot be completely avoided unless we use a unit gain, which will not give any noise reduction; but one can manage to minimize the effect by applying a proper windowing function such as the Kaiser one before FFT and after the IFFT.) With overlap frames, the STFT coefficients from neighboring frames are not independent and there is some correlation among them in principle. To check this, we take a segment of speech from the signal shown in Fig. 1.1, which is 1 second long. We divide this signal into overlap frames with a frame length of 16 ms and 75% overlap. Each frame is then transformed into the STFT domain using a 128-point FFT. For each subband, we compute the cross-correlation coefficients of the complex STFT coefficients. Figure 1.3 plots the magnitude of the cross-correlation

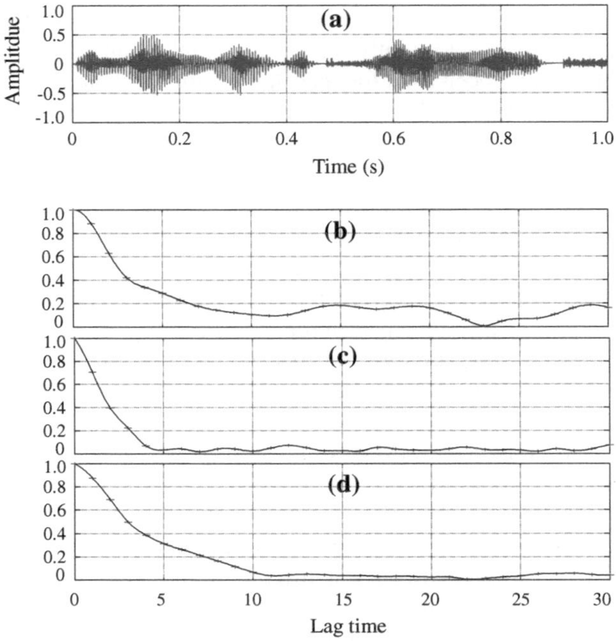

Fig. 1.3 Illustration of the interframe correlation. **a** A 1-second segment of speech signal taken from the signal shown in Fig. 1.1, **b** the magnitude of the interframe cross-correlation coefficients for the 4th frequency bin, **c** the magnitude of the interframe cross-correlation coefficients for the 8th frequency bin, and **d** the magnitude of the interframe cross-correlation coefficients for the 16th frequency bin. The sampling rate is 8 kHz, the frame length is 16 ms (128 points), the FFT size is 128, and the overlapp is 75%

coefficients for the 4th, 8th, and 16th subbands. It is clearly seen that there exists correlation between STFT coefficients from neighboring frames for all the three bands even though the degree of correlation may vary from one band to another. Let us setup a threshold, say 0.2. If the magnitude of the cross-correlation coefficient is smaller than this threshold, the interframe correlation is considered negligible while it is not negligible if the magnitude is larger than the threshold. One can see from Fig. 1.3 that for the 4th and 16th subbands the correlation within six neighboring frames is large and not negligible while for the 8th subband, the correlation cannot be neglected within three neighboring frames. Of course, the number of frames within which the STFT coefficients are correlated depends on how much overlap is used. Figure 1.4 shows the magnitude of the interframe cross-correlation coefficients for the 4th frequency bin when there are, respectively, no, 50%, 75%, and 87.5% overlap between frames next to each other. It is clearly seen that the interframe correlation increases with the amount of overlap. When 87.5% overlap is used, we see that more than 10 neighboring frames are correlated. Even with no overlap, there is still some correlation (magnitude larger than 0.2) between two frames next to each

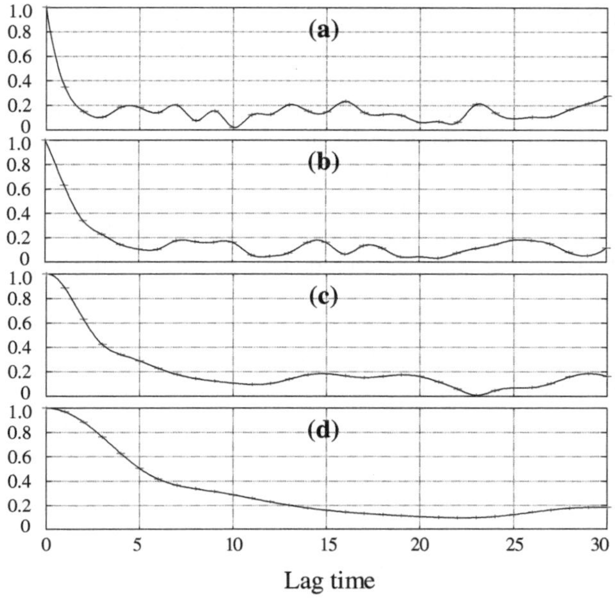

Fig. 1.4 Illustration of the interframe correlation. The magnitude of the interframe cross-correlation coefficients for the 4th frequency bin when **a** there is no overlap, **b** there is 50% overlap, **c** there is 75% overlap, and **d** there is 87.5% overlap. Conditions same as in Fig. 1.3

other. This is because speech signals are self correlated as well proven from the LPC theory [19].

Now that we have shown that there exists correlation between STFT coefficients from neighboring frames, one important question arises: how can we use such interframe correlation for noise reduction? As a matter of fact, some recent efforts have already started to explore this [20, 21]. A comprehensive coverage on how to use the interframe correlation to improve speech enhancement will be discussed in Chap. 3 and Chap. 5 of this work.

1.3 Benefit of Using Multiple Microphones

Traditionally in single-channel noise reduction, speech distortion is inevitable and the amount of speech distortion is in general proportional to the amount of noise reduction [22]. One way to avoid speech distortion is through the use of multiple microphones, leading to an approach called multichannel noise reduction.

The idea of multichannel noise reduction was inspired by the traditional theory of beamforming that dates back to the mid twentieth century and was initially developed for sonar and radar applications [23–25]. The basic principle of beamforming can be described as synchronizing-and-adding. If we assume that the acoustic channels are free of reverberation, the signal components across all sensors can be synchronized by delaying (or advancing) each microphone output by a proper amount of time. When these aligned signals are weighted and summed together, the signal components will be combined coherently and hence reinforced. In contrast, the noise signals are added up incoherently (in power) due to their random nature. This results in a gain for the SNR. However, this simple idea of synchronizing-and-adding, well known as the delay-and-sum beamformer, is good only for narrowband signals because phase delay is frequency dependent. For broadband speech, the directivity pattern of a delay-and-sum beamformer would not be the same across a broad frequency band. If we use such a beamformer, when the steering direction is different from the source incident angle, the source signal will be lowpass filtered. In addition, noise coming from a direction different from the beamformers look direction will not be uniformly attenuated over its entire spectrum. This "spectral tilt" results in a disturbing artifact in the array output [26]. One way to overcome this problem is to perform narrowband decomposition and design narrowband beamformers independently at each frequency. This structure is equivalent to applying a finite-impulse-response (FIR) filter to each microphone output and then summing the filtered signals together. Therefore, this method is often referred to as filter-and-sum beamforming, which was first introduced by Frost [27]. Traditionally, the filter coefficients for a filter-and-sum beamformer are determined based on a prespecified beampattern and hence are independent of the signal characteristics and room reverberation condition. This so-called fixed beamforming method performs reasonably well in anechoic situations where the speech component observed at each microphone is purely a delayed and attenuated copy of the source signal. However, its performance degrades significantly in practical acoustic environments where reverberation is inevitable. One way to improve noise reduction performance in the presence of reverberation is to compute the filter coefficients in an adaptive way based on the room propagation condition. For example, if we know (or can estimate) the signal incident angle, we can optimize the filter coefficients and steer the beamformers look direction such that the desired signal is passed through without attenuation while the signal contributions from all other directions are minimized [23]. This so-called minimum variance distortionless response (MVDR) or Capon method can dramatically improve the beamformers noise reduction performance. However, the speech distortion with this method is also substantial in real acoustic environments where there is reverberation [28]. In order to minimize speech distortion, more sophisticated adaptive algorithms such as linearly constrained minimum variance (LCMV) [27–32], generalized sidelobe canceller (GSC) [33, 34], and multiple-input/output inverse theorem (MINT) [35] were developed. These approaches use the acoustic channel impulse responses from the desired source to the multiple microphones to determine the beamforming filter coefficients. They can achieve high performance when the channel impulse responses are known *a priori* (or can be estimated accurately) and the background noise level is low. However, the performance is very sensitive to the measurement error of channel impulse responses and a small amount of measurement error can lead to significant performance degradation.

Note that beamforming aims at estimating the original source signal. So, unlike the single-channel methods, which exclusively focus on noise reduction, beamforming actually tries to solve both speech dereverberation and noise reduction at the same time. However, speech dereverberation alone is a very difficult task, and there have not been any robust, practical solutions so far. If we consider both dereverberation and noise reduction at the same time, this would only make the problem more complicated. Recently, much efforts have been made to reformulate the beamforming problem to estimate the speech components received at the microphone sensors instead of the source signal [36–38], leading to a new approach called multichannel noise reduction. Multichannel noise reduction ignores speech dereverberation. As a result, it should be much more effective than beamforming in suppressing noise.

Similar to the single-channel case, multichannel noise reduction can be formulated in either the time or frequency domain; but often the STFT domain is preferred for implementation efficiency. In the STFT domain, the basic problem of multichannel noise reduction is how to design either optimal spatial gains or optimal spatio-temporal filters, which will be comprehensively covered in this work.

1.4 Interband Correlation

In the STFT domain, speech enhancement filters are generally a function of the noisy and noise spectra. The two spectra are not known *a priori* and have to be estimated in real applications. A *de facto* standard practice in the field of speech enhancement is to treat the short-time FFT spectrum as an estimate of the true spectrum. Such an estimate, however, generally has very large variations about the true spectrum, causing the estimated gains exceed their theoretical range between 0 and 1. As a result, a nonlinear rectification process has to be used to force the gain to be between 0 and 1. But this would produce some isolated narrowband frequency components in the filtered spectrum. When transformed into the time domain, these isolated components produce music tone sounding noise, which is widely referred to as "musical noise." Musical noise is very unpleasant to hear. Much evidence has shown that listeners would rather prefer to listen to the original noisy signal instead of the enhanced signal with musical noise in most cases. Therefore, it is important not to introduce such noise when we implement a frequency-domain algorithm. One way to limit the amount of musical noise is to smooth the spectral estimates by averaging them over frequency bands [39, 40]. This *ad hoc* trick, though having no theoretical basis, was proven to be effective. Now, an important question one would ask: why interband correlation information is helpful? To answer this question, let us first investigate if there is any correlation between STFT coefficients from different frequency bands. In principle, STFT coefficients from different bands should be orthogonal if the FFT length is sufficiently large. In practice, however, FFT length cannot be very large due to practical reasons such as delay limit, quasi-stationary assumption, etc. With a short FFT length, there will be spectral leakage into neighboring bands, causing interband correlation. To illustrate on this, we use the signal

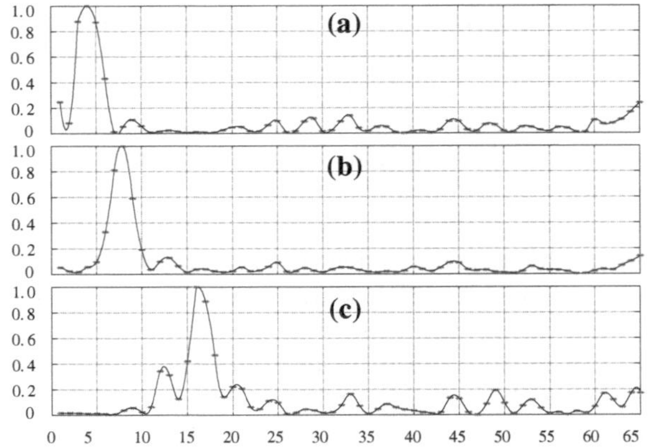

Fig. 1.5 Illustration of the interband correlation. **a** The magnitude of the cross-correlation coefficients between the 4th and other bands, **b** the magnitude of the cross-correlation coefficients between the 8th and other bands, and **c** the magnitude of the cross-correlation coefficients between the 16th and other bands. Conditions same as in Fig. 1.3

shown in Fig. 1.3. Again, we divide the signal into overlap frames with a frame length of 16 ms and 75% overlap. Each frame is transformed into the STFT domain using a 128-point FFT. We then compute the cross-correlation coefficients between different bands. Figure 1.5 plots the results for the 4th, 8th, and 16th frequency bands. It is clear that there is strong correlation between frequency bands that are next to each other. It is this interband correlation that helps reduce musical noise in the aforementioned spectral smoothing technique. Another interesting question then arises: what is the optimal way to use the interband correlation information? This will be addressed in Chap. 6 of this work in the framework of the bifrequency spectrum. The use of the bifrequency spectrum will not only help reduce musical noise, but improve speech enhancement performance as well.

1.5 Organization of the Work

Although noise reduction can be achieved in the STFT domain by using a gain in each frame and subband based on a single-channel observation as discussed in the previous sections, it is helpful to improve performance by using multichannel observations and exploiting interframe and interband correlation. This work addresses the general problem of speech enhancement in the STFT domain and exploits different possible ways in improving noise reduction, including using multichannel, interframe, and interband information. We divide the general problem into five basic categories depending on the number of channels being used and whether interframe or interband correlation is considered. Each category will be covered and discussed in a separate chapter. So, the rest of this work is organized into six chapters.

Chapter 2 investigates the single-channel problem where STFT coefficients at different frames and frequency bands are assumed to be independent. In this case, the noise reduction filter in each frequency band is basically a gain. We discuss how to derive different optimal gains including the well-known ones and how to evaluate them. Chapter 3 also deals with the single-channel problem. The difference between Chap. 3 and Chap. 2 is that now the interframe correlation is taken into account and a filter is applied in each subband instead of just a gain. Chapter 4 considers the problem of multichannel noise reduction in the STFT domain so that the spatial information picked up by an array of microphones can be used to mitigate the noise effect. Like in Chap. 2, the interframe correlation is neglected in this chapter. So, the basic problem is how to derive optimal gains. Chapter 5 addresses the multichannel noise reduction in the STFT domain with interframe correlation being taken into account. In Chap. 6, we consider the interband correlation in the design of noise reduction filters. We illustrate the basic principle for the single-channel case as an example, while this concept can be generalized to other scenarios. Finally, Chap. 7 concludes our work.

References

1. N. Wiener, *Extrapolation, Interpolation, and Smoothing of Stationary Time Series* (Wiley, New York, 1949)
2. M.R. Schroeder, Apparatus for suppressing noise and distortion in communication signals, U.S. Patent No. 3,180,936, filed 1 Dec 1960, issued 27 Apr (1965)
3. M.R. Schroeder, Processing of communication signals to reduce effects of noise, U.S. Patent No. 3,403,224, filed 28 May 1965, issued 24 Sept (1968)
4. S.F. Boll, Suppression of acoustic noise in speech using spectral subtraction. IEEE Trans. Acoust. Speech Signal Process. **ASSP-27**, 113–120 (1979)
5. R.J. McAulay, M.L. Malpass, Speech enhancement using a soft-decision noise suppression filter. IEEE Trans. Acoust. Speech Signal Process. **ASSP-28**, 137–145 (1980)
6. J.S. Lim, A.V. Oppenheim, Enhancement and bandwidth compression of noisy speech. Proc. IEEE **67**, 1586–1604 (1979)
7. M.M. Sondhi, C.E. Schmidt, L.R. Rabiner, Improving the quality of a noisy speech signal. Bell Syst. Techn. J. **60**, 1847–1859 (1981)
8. Y. Ephraim, D. Malah, Speech enhancement using a minimum mean-square error short-time spectral amplitude estimator. IEEE Trans. Acoust. Speech Signal Process. **ASSP-32,** 1109–1121 (1984)
9. Y. Ephraim, D. Malah, Speech enhancement using a minimum mean-square error log-spectral amplitude estimator. IEEE Trans. Acoust. Speech Signal Process. **ASSP-33,** 443–445 (1985)
10. P.J. Wolfe, S.J. Godsill, Simple alternatives to the Ephraim and Malah suppression rule for speech enhancement, in *Proceedings IEEE ICASSP*, pp. 496–499 (2001)
11. L.E. Baum, T. Petrie, Statistical inference for probabilistic functons of finite state Markov chains. Ann. Math. Stat. **73**, 1554–1563 (1966)
12. J.K. Baker, The dragon system–an overview. IEEE Trans. Acoust. Speech Signal Process. **ASSP-23**, 24–29 (1975)

13. F. Jelinek, Continuous speech recognition by statistical methods. Proc. IEEE **64**, 532–536 (1976)
14. L.R. Rabiner, A tutorial on hidden Markov models and selected applications in speech recognition. Proc. IEEE **77**, 257–286 (1989)
15. Y. Ephraim, D. Malah, B.H. Juang, On the application of hidden Markov models for enhancing noisy speech. IEEE Trans. Acoust. Speech Signal Process. **37**, 1846–1856 (1989)
16. Y. Ephraim, A Bayesian estimation approach for speech enhancement using hidden Markov models. IEEE Trans. Signal Process. **40**, 725–735 (1992)
17. Y. Ephraim, Statstical-model-based speech enhancement systems. Proc. IEEE **80**, 1526–1555 (1992)
18. H. Sameti, H. Sheikhzadeh, L. Deng, R.L. Brennan, HMM-based strategies for enhancement of speech signals embedded in nonstationary noise. IEEE Trans. Speech Audio Process. **6**, 445–455 (1998)
19. L.R. Rabiner, B.H. Juang, *Fundamentals of Speech Recognition* (Prentice-Hall, Englewood Cliffs, 1993)
20. J. Benesty, Y. Huang, A single-channel noise reduction MVDR filter, in *Proceedings IEEE ICASSP*, pp. 273–276 (2011)
21. S. So, K.K. Paliwal, Modulation-domain Kalman filtering for single-channel speech enhancemen. Speech Commun. **53**, 818–829 (2011)
22. J. Chen, J. Benesty, Y. Huang, S. Doclo, New insights into the noise reduction Wiener filter. IEEE Trans. Audio Speech Lang. Process. **14**, 1218–1234 (2006)
23. J. Capon, High resolution frequency-wavenumber spectrum analysis. Proc. IEEE **57**, 1408–1418 (1969)
24. L.J. Griffiths, C.W. Jim, An alternative approach to linearly constrained adaptive beamforming. IEEE Trans. Antennas Propag. **AP-30**, 27–34 (1982)
25. B.D. Van Veen, K.M. Buckley, Beamforming: a versatile approach to spatial filtering. IEEE Audio Speech Signal Process Mag. **5**, 4–24 (1988)
26. D.B. Ward, R.C. Williamson, R.A. Kennedy, Broadband microphone arrays for speech acquisition. Acoust. Australia **26**, 17–20 (1998)
27. O.L. Frost, An algorithm for linearly constrained adaptive array processing. Proc. IEEE **60**, 926–935 (1972)
28. J. Benesty, J. Chen, Y. Huang, J. Dmochowski, On microphone array beamforming from a MIMO acoustic signal processing perspective. IEEE Trans. Audio Speech Lang. Process. **15**, 1053–1065 (2008)
29. H. Cox, R. Zeskind, M. Owen, Adaptive beamforming. IEEE Trans. Acoust. Speech Signal Process. **35**, 1365–1376 (1987)
30. M. Brandstein, D.B. Ward (eds.), *Microphone Arrays: Signal Processing Techniques and Applications* (Springer, Berlin, 2001)
31. C.W. Jim, A comparison of two LMS constrained optimal array structures. Proc. IEEE **65**, 1730–1731 (1977)
32. W. Herbordt, W. Kellermann, Adaptive beamforming for audio signal acquisition. in *Adaptive Signal Processing: Applications to Real-World Problems*, ed. by J. Benesty, Y. Huang (Springer, Berlin, 2003), chap. 6, pp. 155–194
33. K.M. Buckley, Broad-band beamforming and the generalized sidelobe canceller. IEEE Trans. Acoust. Speech Signal Process **ASSP-34**, 1322–1323 (1986)
34. S. Werner, J.A. Apolinario, M.L.R. de Campos, On the equivalence of RLS implementations of LCMV and GSC processors. IEEE Signal Process Lett. **10**, 356–359 (2003)
35. M. Miyoshi, Y. Kaneda, Inverse filtering of room acoustics. IEEE Trans. Acoust. Speech Signal Process **36**, 145–152 (1988)
36. S. Doclo, M. Moonen, GSVD-based optimal filtering for single and multimicrophone speech enhancement. IEEE Trans. Signal Process. **50**, 2230–2244 (2002)
37. S. Gannot, I. Cohen, Speech enhancement based on the general transfer function GSC and postfiltering. IEEE Trans. Speech Audio Process. **12**, 561–571 (2004)

38. J. Chen, J. Benesty, Y. Huang, A minimum distortion noise reduction algorithm with multiple microphones. IEEE Trans. Audio Speech Lang. Process. **16**, 481–493 (2008)
39. E.J. Diethorn, A subband noise-reduction method for enhancing speech in telephony and teleconferencing, in *Proceedings IEEE WASPAA* (1977)
40. J. Chen, Y. Huang, J. Benesty, Filtering techniques for noise reduction and speech enhancement, in *Adaptive Signal Processing: Application to Real-World Problems*, ed. by J. Benesty, Y. Huang (Springer, Berlin, 2003) pp. 129–154

Chapter 2
Single-Channel Speech Enhancement with a Gain

There are different ways to perform speech enhancement in the frequency domain from a single microphone signal. The simplest way is to estimate the desired signal from the noisy observation with a simple complex gain. This approach is investigated in this chapter and all well-known optimal gains are derived. We start by explaining the single-channel signal model for speech enhancement in the time and frequency domains.

2.1 Signal Model

The noise reduction or speech enhancement problem considered in this study is one of recovering the desired signal (or clean speech) $x(t)$, t being the time index, of zero mean from the noisy observation (microphone signal) [1–3]

$$y(t) = x(t) + v(t), \tag{2.1}$$

where $v(t)$ is the unwanted additive noise, which is assumed to be a zero-mean random process white or colored but uncorrelated with $x(t)$. All signals are considered to be real and broadband.

Using the short-time Fourier transform (STFT),[1] (2.1) can be rewritten in the frequency domain as

$$Y(k, m) = X(k, m) + V(k, m), \tag{2.2}$$

where the zero-mean complex random variables $Y(k, m)$, $X(k, m)$, and $V(k, m)$ are the STFTs of $y(t)$, $x(t)$, and $v(t)$, respectively, at frequency-bin $k \in \{0, 1, \ldots, K-1\}$ and time-frame m. Since $x(t)$ and $v(t)$ are uncorrelated by assumption, the variance of $Y(k, m)$ is

---

[1] Note that the concepts presented in this work can be applied to any other transformed domain.

J. Benesty et al., *Speech Enhancement in the STFT Domain*, SpringerBriefs in Electrical and Computer Engineering,
DOI: 10.1007/978-3-642-23250-3_2,

$$\begin{aligned}\phi_Y(k,m) &= E\left[|Y(k,m)|^2\right] \\ &= \phi_X(k,m) + \phi_V(k,m),\end{aligned} \tag{2.3}$$

where $E[\cdot]$ denotes mathematical expectation, and

$$\phi_X(k,m) = E\left[|X(k,m)|^2\right], \tag{2.4}$$

$$\phi_V(k,m) = E\left[|V(k,m)|^2\right], \tag{2.5}$$

are the variances of $X(k,m)$ and $V(k,m)$, respectively.

2.2 Microphone Signal Processing with a Gain

In this chapter, we try to estimate the desired signal, $X(k,m)$, from the noisy observation, $Y(k,m)$, i.e.,

$$Z(k,m) = H(k,m)Y(k,m), \tag{2.6}$$

where $Z(k,m)$ is supposed to be the estimate of $X(k,m)$ and $H(k,m)$ is a complex gain that needs to be determined. This procedure is called the single-channel speech enhancement in the STFT domain with a complex gain.

We can express (2.6) as

$$\begin{aligned}Z(k,m) &= H(k,m)\left[X(k,m) + V(k,m)\right] \\ &= X_{\text{fd}}(k,m) + V_{\text{rn}}(k,m),\end{aligned} \tag{2.7}$$

where

$$X_{\text{fd}}(k,m) = H(k,m)X(k,m) \tag{2.8}$$

is the filtered desired signal and

$$V_{\text{rn}}(k,m) = H(k,m)V(k,m) \tag{2.9}$$

is the residual noise.

Since the estimate of the desired signal is the sum of two terms that are uncorrelated, the variance of $Z(k,m)$ is

$$\begin{aligned}\phi_Z(k,m) &= |H(k,m)|^2\,\phi_Y(k,m) \\ &= \phi_{X_{\text{fd}}}(k,m) + \phi_{V_{\text{rn}}}(k,m),\end{aligned} \tag{2.10}$$

where

$$\phi_{X_{\mathrm{fd}}}(k, m) = |H(k, m)|^2 \phi_X(k, m), \tag{2.11}$$

$$\phi_{V_{\mathrm{rn}}}(k, m) = |H(k, m)|^2 \phi_V(k, m), \tag{2.12}$$

are the variances of $X_{\mathrm{fd}}(k, m)$ and $V_{\mathrm{rn}}(k, m)$, respectively.

2.3 Performance Measures

The first attempts to derive relevant and rigorous performance measures in the context of speech enhancement can be found in [1, 4, 5]. These references are the main inspiration for the derivation of measures in the studied context throughout this work.

In this section, we are going to define the most useful performance measures for single-channel speech enhancement with a gain in the STFT domain. We can divide these measures into two categories. The first category evaluates the noise reduction performance while the second one evaluates speech distortion. We are also going to discuss the very convenient mean-square error (MSE) criterion and show how it is related to the performance measures.

2.3.1 *Noise Reduction*

One of the most fundamental measures in all aspects of speech enhancement is the signal-to-noise ratio (SNR). The input SNR is a second-order measure which quantifies the level of noise present relative to the level of the desired signal.

We define the subband and fullband input SNRs at time-frame m as [1]

$$\mathrm{iSNR}(k, m) = \frac{\phi_X(k, m)}{\phi_V(k, m)}, \quad k = 0, 1, \ldots, K - 1, \tag{2.13}$$

$$\mathrm{iSNR}(m) = \frac{\sum_{k=0}^{K-1} \phi_X(k, m)}{\sum_{k=0}^{K-1} \phi_V(k, m)}. \tag{2.14}$$

It is easy to show that [1]

$$\mathrm{iSNR}(m) \leq \sum_{k=0}^{K-1} \mathrm{iSNR}(k, m). \tag{2.15}$$

To quantify the level of the noise remaining after the noise reduction processing via the complex gain, we define the output SNR as the ratio of the variance of the filtered desired signal over the variance of the residual noise. We easily deduce the subband output SNR

$$\begin{aligned} \text{oSNR}\left[H(k,m)\right] &= \frac{|H(k,m)|^2\,\phi_X(k,m)}{|H(k,m)|^2\,\phi_V(k,m)} \\ &= \frac{\phi_X(k,m)}{\phi_V(k,m)},\ k = 0, 1, \ldots, K-1 \end{aligned} \tag{2.16}$$

and the fullband output SNR

$$\text{oSNR}\left[H(:,m)\right] = \frac{\sum_{k=0}^{K-1} |H(k,m)|^2\,\phi_X(k,m)}{\sum_{k=0}^{K-1} |H(k,m)|^2\,\phi_V(k,m)}. \tag{2.17}$$

We notice that the subband output SNR is equal to the subband input SNR, so the subband SNR cannot be improved with just a gain but the fullband output SNR can. It can be verified that [1]

$$\text{oSNR}\left[H(:,m)\right] \le \sum_{k=0}^{K-1} \text{iSNR}(k,m). \tag{2.18}$$

The previous inequality shows that the fullband output SNR is always upper bounded no matter the choices of the $H(k, m)$.

For the particular gain $H(k,m) = 1$, we have

$$\text{oSNR}\left[1(k,m)\right] = \text{iSNR}(k,m),\ k = 0, 1, \ldots, K-1, \tag{2.19}$$

$$\text{oSNR}\left[1(:,m)\right] = \text{iSNR}(m). \tag{2.20}$$

With the identity gain, 1, the SNR cannot be improved.

The noise reduction factor [4, 5] quantifies the amount of noise whose is rejected by the complex gain. This quantity is defined as the ratio of the variance of the noise at the microphone over the variance of the residual noise. The subband and fullband noise reduction factors are then

$$\begin{aligned} \xi_{\text{nr}}\left[H(k,m)\right] &= \frac{\phi_V(k,m)}{|H(k,m)|^2\,\phi_V(k,m)} \\ &= \frac{1}{|H(k,m)|^2},\ k = 0, 1, \ldots, K-1, \end{aligned} \tag{2.21}$$

$$\begin{aligned} \xi_{\text{nr}}\left[H(:,m)\right] &= \frac{\sum_{k=0}^{K-1} \phi_V(k,m)}{\sum_{k=0}^{K-1} |H(k,m)|^2\,\phi_V(k,m)} \\ &= \frac{\sum_{k=0}^{K-1} \phi_V(k,m)}{\sum_{k=0}^{K-1} \xi_{\text{nr}}^{-1}\left[H(k,m)\right]\phi_V(k,m)}, \end{aligned} \tag{2.22}$$

and we always have

$$\xi_{\rm nr}\left[H(:,m)\right] \leq \sum_{k=0}^{K-1} \xi_{\rm nr}\left[H(k,m)\right]. \tag{2.23}$$

The noise reduction factors are expected to be lower bounded by 1 for appropriate choices of the $H(k, m)$. So the more the noise is reduced, the higher are the values of the noise reduction factors.

2.3.2 Speech Distortion

In practice, the complex gain distorts the desired signal. In order to evaluate the level of this distortion, we define the speech reduction factor [1] as the variance of the desired signal over the variance of the filtered desired signal. Therefore, the subband and fullband speech reduction factors are defined as

$$\begin{aligned}
\xi_{\rm sr}\left[H(k,m)\right] &= \frac{\phi_X(k,m)}{|H(k,m)|^2\,\phi_X(k,m)} \\
&= \frac{1}{|H(k,m)|^2},\quad k=0,1,\ldots,K-1,
\end{aligned} \tag{2.24}$$

$$\begin{aligned}
\xi_{\rm sr}\left[H(:,m)\right] &= \frac{\sum_{k=0}^{K-1}\phi_X(k,m)}{\sum_{k=0}^{K-1}|H(k,m)|^2\,\phi_X(k,m)} \\
&= \frac{\sum_{k=0}^{K-1}\phi_X(k,m)}{\sum_{k=0}^{K-1}\xi_{\rm sr}^{-1}\left[H(k,m)\right]\phi_X(k,m)},
\end{aligned} \tag{2.25}$$

and we always have

$$\xi_{\rm sr}\left[H(:,m)\right] \leq \sum_{k=0}^{K-1} \xi_{\rm sr}\left[H(k,m)\right]. \tag{2.26}$$

The speech reduction factor is equal to 1 if there is no distortion and expected to be greater than 1 when distortion occurs.

By making the appropriate substitutions, one can derive the relationships:

$$\frac{\mathrm{oSNR}\left[H(k,m)\right]}{\mathrm{iSNR}(k,m)} = \frac{\xi_{\rm nr}\left[H(k,m)\right]}{\xi_{\rm sr}\left[H(k,m)\right]},\quad k=0,1,\ldots,K-1, \tag{2.27}$$

$$\frac{\mathrm{oSNR}\left[H(:,m)\right]}{\mathrm{iSNR}(m)} = \frac{\xi_{\rm nr}\left[H(:,m)\right]}{\xi_{\rm sr}\left[H(:,m)\right]}. \tag{2.28}$$

These expressions indicate the equivalence between gain/loss in SNR and distortion for both the subband and fullband cases.

Another way to measure the distortion of the desired speech signal due to the complex gain is the speech distortion index [1, 4, 5], which is defined as the mean-square error between the desired signal and the filtered desired signal, normalized by the variance of the desired signal, i.e.,

$$\begin{aligned} \upsilon_{\mathrm{sd}}\left[H(k,m)\right] &= \frac{E\left\{|H(k,m)X(k,m) - X(k,m)|^2\right\}}{\phi_X(k,m)} \\ &= |H(k,m) - 1|^2, \ k = 0, 1, \ldots, K-1 \end{aligned} \tag{2.29}$$

in the subband case and

$$\begin{aligned} \upsilon_{\mathrm{sd}}\left[H(:,m)\right] &= \frac{\sum_{k=0}^{K-1} E\left\{|H(k,m)X(k,m) - X(k,m)|^2\right\}}{\sum_{k=0}^{K-1} \phi_X(k,m)} \\ &= \frac{\sum_{k=0}^{K-1} \upsilon_{\mathrm{sd}}\left[H(k,m)\right]\phi_X(k,m)}{\sum_{k=0}^{K-1} \phi_X(k,m)} \end{aligned} \tag{2.30}$$

in the fullband case. It can be verified that

$$\upsilon_{\mathrm{sd}}\left[H(:,m)\right] \leq \sum_{k=0}^{K-1} \upsilon_{\mathrm{sd}}\left[H(k,m)\right]. \tag{2.31}$$

However, the speech distortion indices are usually upper bounded by 1 for optimal gains.

2.3.3 *Mean-Square Error Criterion*

Error criteria play a critical role in deriving optimal gains. The MSE [6] is, by far, the most practical one.

In the STFT domain, the error signal between the estimated and desired signals at the frequency-bin k and time-frame m is

$$\begin{aligned} \mathcal{E}(k,m) &= Z(k,m) - X(k,m) \\ &= H(k,m)Y(k,m) - X(k,m), \end{aligned} \tag{2.32}$$

which can also be written as the sum of two uncorrelated error signals:

$$\mathcal{E}(k,m) = \mathcal{E}_{\mathrm{d}}(k,m) + \mathcal{E}_{\mathrm{r}}(k,m), \tag{2.33}$$

where

$$\mathcal{E}_{\mathrm{d}}(k,m) = \left[H(k,m) - 1\right]X(k,m) \tag{2.34}$$

is the speech distortion due to the gain and

$$\mathcal{E}_{\rm r}(k,m) = H(k,m)V(k,m) \tag{2.35}$$

represents the residual noise.

The subband MSE criterion is then

$$\begin{aligned} J\left[H(k,m)\right] &= E\left[|\mathcal{E}(k,m)|^2\right] \\ &= \phi_X(k,m) + |H(k,m)|^2\,\phi_Y(k,m) - 2\mathcal{R}\left[H(k,m)\phi_{YX}(k,m)\right] \\ &= \phi_X(k,m) + |H(k,m)|^2\,\phi_Y(k,m) - 2\mathcal{R}\left[H(k,m)\phi_X(k,m)\right] \\ &= J_{\rm d}\left(k,m\right) + J_{\rm r}\left(k,m\right), \end{aligned} \tag{2.36}$$

where $\mathcal{R}[\cdot]$ is the real part of a complex number,

$$\begin{aligned} \phi_{YX}(k,m) &= E\left[Y(k,m)X^*(k,m)\right] \\ &= \phi_X(k,m) \end{aligned}$$

is the cross-correlation between the signals $Y(k,m)$ and $X(k,m)$, superscript * denotes complex conjugation,

$$\begin{aligned} J_{\rm d}\left[H(k,m)\right] &= E\left[|\mathcal{E}_{\rm d}(k,m)|^2\right] \\ &= |H(k,m) - 1|^2\,\phi_X(k,m) \\ &= \upsilon_{\rm sd}\left[H(k,m)\right]\phi_X(k,m), \end{aligned} \tag{2.37}$$

and

$$\begin{aligned} J_{\rm r}\left[H(k,m)\right] &= E\left[|\mathcal{E}_{\rm r}(k,m)|^2\right] \\ &= |H(k,m)|^2\,\phi_V(k,m) \\ &= \frac{\phi_V(k,m)}{\xi_{\rm nr}\left[H(k,m)\right]}. \end{aligned} \tag{2.38}$$

Two particular gains are of great interest: $H(k,m) = 1$ and $H(k,m) = 0$. With the first one (identity gain), we have neither noise reduction nor speech distortion and with the second one (zero gain), we have maximum noise reduction and maximum speech distortion. For both gains, however, it can be verified that the output SNR is equal to the input SNR. For these two particular gains, the subband MSEs are

$$J\left[1(k,m)\right] = J_{\rm r}\left[1(k,m)\right] = \phi_V(k,m), \tag{2.39}$$

$$J\left[0(k,m)\right] = J_{\rm d}\left[0(k,m)\right] = \phi_X(k,m). \tag{2.40}$$

As a result,

$$\mathrm{iSNR}(k,m) = \frac{J\left[0(k,m)\right]}{J\left[1(k,m)\right]}. \tag{2.41}$$

We define the subband normalized MSE (NMSE) with respect to $J\left[1(k,m)\right]$ as

$$\begin{aligned}\widetilde{J}\left[H(k,m)\right] &= \frac{J\left[H(k,m)\right]}{J\left[1(k,m)\right]} \\ &= \mathrm{iSNR}(k,m)\cdot \upsilon_{\mathrm{sd}}\left[H(k,m)\right] + \frac{1}{\xi_{\mathrm{nr}}\left[H(k,m)\right]} \\ &= \mathrm{iSNR}(k,m)\left\{\upsilon_{\mathrm{sd}}\left[H(k,m)\right] + \frac{1}{\mathrm{oSNR}\left[H(k,m)\right]\cdot \xi_{\mathrm{sr}}\left[H(k,m)\right]}\right\},\end{aligned} \tag{2.42}$$

where

$$\upsilon_{\mathrm{sd}}\left[H(k,m)\right] = \frac{J_{\mathrm{d}}\left[H(k,m)\right]}{J_{\mathrm{d}}\left[0(k,m)\right]}, \tag{2.43}$$

$$\mathrm{iSNR}(k,m)\cdot \upsilon_{\mathrm{sd}}\left[H(k,m)\right] = \frac{J_{\mathrm{d}}\left[H(k,m)\right]}{J_{\mathrm{r}}\left[1(k,m)\right]}, \tag{2.44}$$

$$\xi_{\mathrm{nr}}\left[H(k,m)\right] = \frac{J_{\mathrm{r}}\left[1(k,m)\right]}{J_{\mathrm{r}}\left[H(k,m)\right]}, \tag{2.45}$$

$$\mathrm{oSNR}\left[H(k,m)\right]\cdot \xi_{\mathrm{sr}}\left[H(k,m)\right] = \frac{J_{\mathrm{d}}\left[0(k,m)\right]}{J_{\mathrm{r}}\left[H(k,m)\right]}. \tag{2.46}$$

This shows how this subband NMSE and the different subband MSEs are related to the performance measures.

We define the subband NMSE with respect to $J\left[0(k,m)\right]$ as

$$\begin{aligned}\overline{J}\left[H(k,m)\right] &= \frac{J\left[H(k,m)\right]}{J\left[0(k,m)\right]} \\ &= \upsilon_{\mathrm{sd}}\left[H(k,m)\right] + \frac{1}{\mathrm{oSNR}\left[H(k,m)\right]\cdot \xi_{\mathrm{sr}}\left[H(k,m)\right]}\end{aligned} \tag{2.47}$$

and, obviously,

$$\widetilde{J}\left[H(k,m)\right] = \mathrm{iSNR}(k,m)\cdot \overline{J}\left[H(k,m)\right]. \tag{2.48}$$

We are only interested in gains for which

$$J_{\mathrm{d}}\left[1(k,m)\right] \le J_{\mathrm{d}}\left[H(k,m)\right] < J_{\mathrm{d}}\left[0(k,m)\right], \tag{2.49}$$

$$J_{\mathrm{r}}\left[0(k,m)\right] < J_{\mathrm{r}}\left[H(k,m)\right] < J_{\mathrm{r}}\left[1(k,m)\right]. \tag{2.50}$$

From the two previous expressions, we deduce that

$$0 \leq \upsilon_{\mathrm{sd}}[H(k,m)] < 1, \tag{2.51}$$

$$1 < \xi_{\mathrm{nr}}[H(k,m)] < \infty. \tag{2.52}$$

It is clear that the objective of noise reduction in the STFT domain is to find optimal gains that would either minimize $J[H(k,m)]$ or minimize $J_{\mathrm{d}}[H(k,m)]$ or $J_{\mathrm{r}}[H(k,m)]$ subject to some constraint.

In the same way, we define the fullband MSE at time-frame m as

$$\begin{aligned} J[H(:,m)] &= \frac{1}{K}\sum_{k=0}^{K-1} J[H(k,m)] \\ &= \frac{1}{K}\sum_{k=0}^{K-1} J_{\mathrm{d}}[H(k,m)] + \frac{1}{K}\sum_{k=0}^{K-1} J_{\mathrm{r}}[H(k,m)] \\ &= J_{\mathrm{d}}[H(:,m)] + J_{\mathrm{r}}[H(:,m)]. \end{aligned} \tag{2.53}$$

We then deduce the fullband NMSEs at time-frame m:

$$\begin{aligned} \widetilde{J}[H(:,m)] &= K\frac{J[H(:,m)]}{\sum_{k=0}^{K-1}\phi_V(k,m)} \\ &= \mathrm{iSNR}(m)\cdot \upsilon_{\mathrm{sd}}[H(:,m)] + \frac{1}{\xi_{\mathrm{nr}}[H(:,m)]}, \end{aligned} \tag{2.54}$$

$$\begin{aligned} \overline{J}[H(:,m)] &= K\frac{J[H(:,m)]}{\sum_{k=0}^{K-1}\phi_X(k,m)} \\ &= \upsilon_{\mathrm{sd}}[H(:,m)] + \frac{1}{\mathrm{oSNR}[H(:,m)]\cdot \xi_{\mathrm{sr}}[H(:,m)]}. \end{aligned} \tag{2.55}$$

It is straightforward to see that minimizing the subband MSE at each frequency-bin k is equivalent to minimizing the fullband MSE.

2.4 Optimal Gains

In this section, we are going to derive the most important gains that can help mitigate the level of the noise picked up by the microphone.

2.4.1 Wiener

By minimizing $J[H(k,m)]$ [Eq. (2.36)] with respect to $H(k,m)$, we easily find the Wiener gain

$$\begin{aligned} H_{\mathrm{W}}(k, m) &= \frac{E\left[|X(k, m)|^2\right]}{E\left[|Y(k, m)|^2\right]} \\ &= 1 - \frac{E\left[|V(k, m)|^2\right]}{E\left[|Y(k, m)|^2\right]} \\ &= \frac{\phi_X(k, m)}{\phi_X(k, m) + \phi_V(k, m)} \\ &= \frac{\mathrm{iSNR}(k, m)}{1 + \mathrm{iSNR}(k, m)}. \end{aligned} \tag{2.56}$$

We see that the noncausal Wiener gain is always real and positive. Furthermore, $0 \leq H_{\mathrm{W}}(k, m) \leq 1, \forall k, m$, and

$$\lim_{\mathrm{iSNR}(k, m) \to \infty} H_{\mathrm{W}}(k, m) = 1, \tag{2.57}$$

$$\lim_{\mathrm{iSNR}(k, m) \to 0} H_{\mathrm{W}}(k, m) = 0. \tag{2.58}$$

We deduce the different subband performance measures:

$$\widetilde{J}\left[H_{\mathrm{W}}(k, m)\right] = \frac{\mathrm{iSNR}(k, m)}{1 + \mathrm{iSNR}(k, m)} \leq 1, \tag{2.59}$$

$$\begin{aligned} \xi_{\mathrm{nr}}\left[H_{\mathrm{W}}(k, m)\right] &= \left[1 + \frac{1}{\mathrm{iSNR}(k, m)}\right]^2 \geq 1 \\ &= \xi_{\mathrm{sr}}\left[H_{\mathrm{W}}(k, m)\right], \end{aligned} \tag{2.60}$$

$$\upsilon_{\mathrm{sd}}\left[H_{\mathrm{W}}(k, m)\right] = \frac{1}{\left[1 + \mathrm{iSNR}(k, m)\right]^2} \leq 1. \tag{2.61}$$

The fullband output SNR is

$$\mathrm{oSNR}\left[H_{\mathrm{W}}(:, m)\right] = \frac{\sum_{k=0}^{K-1} \phi_X(k, m) \left[\dfrac{\mathrm{iSNR}(k, m)}{1 + \mathrm{iSNR}(k, m)}\right]^2}{\sum_{k=0}^{K-1} \phi_V(k, m) \left[\dfrac{\mathrm{iSNR}(k, m)}{1 + \mathrm{iSNR}(k, m)}\right]^2}. \tag{2.62}$$

We observe from the previous expression that if the subband input SNR is constant across frequencies then the fullband SNR cannot be improved.

Property 2.1 *With the optimal STFT-domain Wiener gain given in* (2.56), *the fullband output SNR is always greater than or equal to the fullband input SNR, i.e.,* $\mathrm{oSNR}\left[H(:, m)\right] \geq \mathrm{iSNR}(m)$.

Proof We can use exactly the same techniques as the ones exposed in [1] to show this property.

Property 2.2 *We have*

$$\frac{\text{iSNR}(m)}{1+\text{oSNR}\left[H_{\text{W}}(:,m)\right]} \le \widetilde{J}\left[H_{\text{W}}(:,m)\right] \le \frac{\text{iSNR}(m)}{1+\text{iSNR}(m)}, \tag{2.63}$$

$$\begin{aligned}\frac{\{1+\text{oSNR}\left[H_{\text{W}}(:,m)\right]\}^2}{\text{iSNR}(m)\cdot\text{oSNR}\left[H_{\text{W}}(:,m)\right]} &\le \xi_{\text{nr}}\left[H_{\text{W}}(:,m)\right] \\ &\le \frac{\left[1+\text{iSNR}(m)\right]\{1+\text{oSNR}\left[H_{\text{W}}(:,m)\right]\}}{\text{iSNR}^2(m)},\end{aligned} \tag{2.64}$$

$$\begin{aligned}\frac{1}{\{1+\text{oSNR}\left[H_{\text{W}}(:,m)\right]\}^2} &\le \upsilon_{\text{sd}}\left[H_{\text{W}}(:,m)\right] \\ &\le \frac{1+\text{oSNR}\left[H_{\text{W}}(:,m)\right]-\text{iSNR}(m)}{\left[1+\text{iSNR}(m)\right]\{1+\text{oSNR}\left[H_{\text{W}}(:,m)\right]\}}.\end{aligned} \tag{2.65}$$

Proof We can use exactly the same techniques as the ones exposed in [1] to show these different inequalities.

2.4.2 Tradeoff

The tradeoff gain is obtained by minimizing the speech distortion with the constraint that the residual noise level is equal to a value smaller than the level of the original noise. This is equivalent to solving the problem

$$\min_{H(k,m)} J_{\text{d}}\left[H(k,m)\right] \quad \text{subject to} \quad J_{\text{r}}\left[H(k,m)\right] = \beta\phi_V(k,m), \tag{2.66}$$

where

$$J_{\text{d}}\left[H(k,m)\right] = \left|H(k,m)-1\right|^2 \phi_X(k,m), \tag{2.67}$$

$$J_{\text{r}}\left[H(k,m)\right] = \left|H(k,m)\right|^2 \phi_V(k,m), \tag{2.68}$$

and $0 < \beta < 1$ in order to have some noise reduction at the frequency-bin k. If we use a Lagrange multiplier, $\mu \ge 0$, to adjoin the constraint to the cost function, we get the tradeoff gain

$$\begin{aligned}H_{\text{T},\mu}(k,m) &= \frac{\phi_X(k,m)}{\phi_X(k,m)+\mu\phi_V(k,m)} \\ &= \frac{\phi_Y(k,m)-\phi_V(k,m)}{\phi_Y(k,m)+(\mu-1)\phi_V(k,m)} \\ &= \frac{\text{iSNR}(k,m)}{\mu+\text{iSNR}(k,m)}.\end{aligned} \tag{2.69}$$

This gain can be seen as a STFT-domain Wiener gain with adjustable input noise level $\mu\phi_V(k, m)$. The particular cases of $\mu = 1$ and $\mu = 0$ correspond to the Wiener and distortionless gains, respectively.

The fullband output SNR is

$$\text{oSNR}\left[H_{\text{T},\mu}(:, m)\right] = \frac{\sum_{k=0}^{K-1} \phi_X(k, m) \left[\frac{\text{iSNR}(k, m)}{\mu + \text{iSNR}(k, m)}\right]^2}{\sum_{k=0}^{K-1} \phi_V(k, m) \left[\frac{\text{iSNR}(k, m)}{\mu + \text{iSNR}(k, m)}\right]^2}. \tag{2.70}$$

Property 2.3 *With the STFT-domain tradeoff gain given in* (2.69), *the fullband output SNR is always greater than or equal to the fullband input SNR, i.e.,* $\text{oSNR}\left[H_{\text{T},\mu}(:, m)\right] \geq \text{iSNR}(m), \forall\mu \geq 0$.

Proof We can use exactly the same techniques as the ones exposed in [1] to show this property.

From (2.70), we deduce that

$$\lim_{\mu\to\infty} \text{oSNR}\left[H_{\text{T},\mu}(:, m)\right] = \frac{\sum_{k=0}^{K-1} \phi_X(k, m)\text{iSNR}^2(k, m)}{\sum_{k=0}^{K-1} \phi_V(k, m)\text{iSNR}^2(k, m)} \leq \sum_{k=0}^{K-1} \text{iSNR}(k, m). \tag{2.71}$$

This shows the trend of the fullband output SNR of the tradeoff gain.

The fullband speech distortion index is

$$\upsilon_{\text{sd}}\left[H_{\text{T},\mu}(:, m)\right] = \frac{\sum_{k=0}^{K-1} \frac{\mu^2\phi_X(k, m)}{[\mu + \text{iSNR}(k, m)]^2}}{\sum_{k=0}^{K-1} \phi_X(k, m)}. \tag{2.72}$$

Property 2.4 *The fullband speech distortion index of the STFT-domain tradeoff gain is an increasing function of the parameter* μ.

Proof It is straightforward to verify that

$$\frac{d\upsilon_{\text{sd}}\left[H_{\text{T},\mu}(:, m)\right]}{d\mu} \geq 0, \tag{2.73}$$

which ends the proof.

It is clear that

$$0 \leq \upsilon_{\text{sd}}\left[H_{\text{T},\mu}(:, m)\right] \leq 1, \ \forall\mu \geq 0. \tag{2.74}$$

Therefore, as μ increases, the fullband output SNR increases at the price of more distortion to the desired signal.

The tradeoff gain can be more general if we make the factor β dependent on the frequency, i.e., $\beta(k)$. By doing so, the control between noise reduction and speech distortion can be more effective since each frequency-bin k can be controlled independently of the others. With this consideration, we can easily see that the optimal gain derived from the criterion (2.66) is now

$$H_{\mathrm{T},\mu}(k,m) = \frac{\mathrm{iSNR}(k,m)}{\mu(k) + \mathrm{iSNR}(k,m)}, \tag{2.75}$$

where $\mu(k)$ is the frequency-dependent Lagrange multiplier. This approach can now provide some noise spectral shaping for masking by the speech signal [7–12].

2.4.3 Maximum Signal-to-Noise Ratio

Let us define the $K \times 1$ vector

$$\mathbf{h}(m) = [H(0,m)\ H(1,m) \cdots H(K-1,m)]^T, \tag{2.76}$$

where the superscript ^{T} denotes transpose of a vector or a matrix. The filter $\mathbf{h}(m)$ contains all the subband gains. The fullband output SNR can be rewritten as

$$\begin{aligned} \mathrm{oSNR}\left[H(:,m)\right] &= \mathrm{oSNR}\left[\mathbf{h}(m)\right] \\ &= \frac{\mathbf{h}^H(m)\mathbf{D}_{\phi_X}(m)\mathbf{h}(m)}{\mathbf{h}^H(m)\mathbf{D}_{\phi_V}(m)\mathbf{h}(m)}, \end{aligned} \tag{2.77}$$

where the superscript ^{H} denotes transpose-conjugate and

$$\mathbf{D}_{\phi_X}(m) = \mathrm{diag}\left[\phi_X(0,m), \phi_X(1,m), \ldots, \phi_X(K-1,m)\right], \tag{2.78}$$

$$\mathbf{D}_{\phi_V}(m) = \mathrm{diag}\left[\phi_V(0,m), \phi_V(1,m), \ldots, \phi_V(K-1,m)\right], \tag{2.79}$$

are two diagonal matrices. We assume here that $\phi_V(k,m) \neq 0, \forall k, m$.

In the maximum SNR approach, we find the filter, $\mathbf{h}(m)$, that maximizes the fullband output SNR defined in (2.77). The solution to this problem that we denote by $\mathbf{h}_{\max}(m)$ is simply the eigenvector corresponding to the maximum eigenvalue of the matrix $\mathbf{D}_{\phi_V}^{-1}(m)\mathbf{D}_{\phi_X}(m)$. Since this matrix is diagonal, its maximum eigenvalue is its largest diagonal element, i.e.,

$$\max_k \frac{\phi_X(k,m)}{\phi_V(k,m)} = \max_k \ \mathrm{iSNR}(k,m). \tag{2.80}$$

Assume that this maximum is the k_0th diagonal element of the matrix $\mathbf{D}_{\phi_V}^{-1}(m)\mathbf{D}_{\phi_X}(m)$. In this case, the k_0th component of $\mathbf{h}_{\max}(m)$ is 1 and all its other components are 0. As a result,

$$\begin{aligned} \text{oSNR}\left[\mathbf{h}_{\max}(m)\right] &= \max_k \ \text{iSNR}(k, m) \\ &= \text{iSNR}(k_0, m). \end{aligned} \tag{2.81}$$

We also deduce that

$$\text{oSNR}\left[\mathbf{h}(m)\right] \le \max_k \ \text{iSNR}(k, m), \ \ \forall \mathbf{h}(m). \tag{2.82}$$

This means that with the Wiener, tradeoff, or any other gain, the fullband output SNR cannot exceed the maximum subband input SNR, which is a very interesting result on its own.

It is easy to derive the fullband speech distortion index:

$$\upsilon_{\text{sd}}\left[\mathbf{h}_{\max}(m)\right] = 1 - \frac{\phi_X(k_0, m)}{\sum_{k=0}^{K-1} \phi_X(k, m)}, \tag{2.83}$$

which can be very close to 1, implying very large distortions of the desired signal.

Needless to say that this maximum SNR filter is never used in practice since all subband signals but one are suppressed. But this filter is still interesting from a theoretical point of view.

References

1. J. Benesty, J. Chen, Y. Huang, I. Cohen, *Noise Reduction in Speech Processing* (Springer, Berlin, 2009)
2. P. Loizou, *Speech Enhancement: Theory and Practice* (CRC Press, Boca Raton, 2007)
3. P. Vary, R. Martin, *Digital Speech Transmission: Enhancement, Coding and Error Concealment* (Wiley, Chichester, 2006)
4. J. Benesty, J. Chen, Y. Huang, S. Doclo, Study of the Wiener filter for noise reduction, in *Speech Enhancement*, ed. by J. Benesty, S. Makino, J. Chen (Springer, Berlin, 2005) pp. 9–41
5. J. Chen, J. Benesty, Y. Huang, S. Doclo, New insights into the noise reduction Wiener filter. IEEE Trans. Audio Speech Lang. Process **14**, 1218–1234 (2006)
6. S. Haykin, *Adaptive Filter Theory*. 4th edn. (Prentice-Hall, Upper Saddle River, 2002)
7. Y. Ephraim, H.L. Van Trees, A signal subspace approach for speech enhancement. IEEE Trans. Speech Audio Process **3**, 251–266 (1995)
8. N. Virag, Single channel speech enhancement based on masking properties of the human auditory system. IEEE Trans. Speech Audio Process **7**, 126–137 (1999)
9. R. Vetter, Single channel speech enhancement using MDL-based subspace approach in Bark domain, in *Proceedings IEEE ICASSP*, pp. 641–644 (2001)
10. Y. Hu, P.C. Loizou, A generalized subspace approach for enhancing speech corrupted by colored noise. IEEE Trans. Speech Audio Process **11**, 334–341 (2003)
11. Y. Hu, P.C. Loizou, A perceptually motivated approach for speech enhancement. IEEE Trans. Speech Audio Process **11**, 457–465 (2003)
12. F. Jabloun, B. Champagne, Incorporating the human hearing properties in the signal subspace approach for speech enhancement. IEEE Trans. Speech Audio Process **11**, 700–708 (2003)

Chapter 3
Single-Channel Speech Enhancement with a Filter

In the previous chapter, we estimated the desired signal from the current data frame only. However, as shown in Chap. 1, consecutive data frames are correlated and this information should, therefore, be taken into account in order to enhance the estimation of the speech signal. Now, a filter needs to be used instead of a gain to reflect this new situation. This chapter, based on the ideas proposed in [1, 2], investigates this framework but we are still concerned with the single-microphone case; therefore, the signal model is the same as in Chap. 2, Sect. 2.1. We start by explaining the principle of linear filtering in this context.

3.1 Microphone Signal Processing with a Filter

Since the interframe correlation is taken into account, we estimate $X(k,m)$, $k = 0, 1, \ldots, K-1$, by passing $Y(k,m)$, $k = 0, 1, \ldots, K-1$, from consecutive time-frames through a finite-impulse-response (FIR) filter of length L, i.e.,

$$\begin{aligned} Z(k,m) &= \sum_{l=0}^{L-1} H_l(k,m) Y(k, m-l) \\ &= \mathbf{h}^H(k,m)\mathbf{y}(k,m), \quad k = 0, 1, \ldots, K-1, \end{aligned} \tag{3.1}$$

where L is the number of consecutive time-frames and

$$\begin{aligned} \mathbf{h}(k,m) &= \left[H_0^*(k,m) \; H_1^*(k,m) \cdots H_{L-1}^*(k,m) \right]^T, \\ \mathbf{y}(k,m) &= \left[Y(k,m) \; Y(k,m-1) \cdots Y(k,m-L+1) \right]^T, \end{aligned}$$

are vectors of length L. The case $L = 1$ corresponds to the conventional STFT-domain approach (see Chap. 2). Note that this concept, to take into account the interframe correlation in a speech enhancement algorithm, was introduced in [3–5] but in the Karhunen-Loève expansion (KLE) domain. In [6], the interframe correlation was also used to improve the a priori SNR estimator.

J. Benesty et al., *Speech Enhancement in the STFT Domain*,
SpringerBriefs in Electrical and Computer Engineering,
DOI: 10.1007/978-3-642-23250-3_3,

Let us now rewrite the signal $Z(k, m)$ into the following form:

$$\begin{aligned} Z(k, m) &= \mathbf{h}^H(k, m)\mathbf{x}(k, m) + \mathbf{h}^H(k, m)\mathbf{v}(k, m) \\ &= X_{\mathrm{f}}(k, m) + V_{\mathrm{rn}}(k, m), \quad k = 0, 1, \ldots, K - 1, \end{aligned} \tag{3.2}$$

where

$$X_{\mathrm{f}}(k, m) = \mathbf{h}^H(k, m)\mathbf{x}(k, m) \tag{3.3}$$

is a filtered version of the desired signal at L consecutive time-frames,

$$V_{\mathrm{rn}}(k, m) = \mathbf{h}^H(k, m)\mathbf{v}(k, m) \tag{3.4}$$

is the residual noise which is uncorrelated with $X_{\mathrm{f}}(k, m)$, and $\mathbf{x}(k, m)$ and $\mathbf{v}(k, m)$ are defined in a similar way to $\mathbf{y}(k, m)$.

At time-frame m, our desired signal is $X(k, m)$ [and not the whole vector $\mathbf{x}(k, m)$]. However, the vector $\mathbf{x}(k, m)$ in $X_{\mathrm{f}}(k, m)$ [Eq. (3.3)] contains both the desired signal, $X(k, m)$, and the components $X(k, m - l)$, $l \neq 0$, which are not the desired signals at time-frame m but signals that are correlated with $X(k, m)$. Therefore, the elements $X(k, m - l)$, $l \neq 0$, contain both a part of the desired signal and a component that we consider as an interference. This suggests that we should decompose $X(k, m - l)$ into two orthogonal components corresponding to the part of the desired signal and interference, i.e.,

$$X(k, m - l) = \rho_X^*(k, m, l)X(k, m) + X_{\mathrm{i}}(k, m - l), \tag{3.5}$$

where

$$X_{\mathrm{i}}(k, m - l) = X(k, m - l) - \rho_X^*(k, m, l)X(k, m), \tag{3.6}$$

$$E[X(k, m)X_{\mathrm{i}}^*(k, m - l)] = 0, \tag{3.7}$$

and

$$\rho_X(k, m, l) = \frac{E\left[X(k, m)X^*(k, m - l)\right]}{E\left[|X(k, m)|^2\right]} \tag{3.8}$$

is the interframe correlation coefficient of the signal $X(k, m)$. Hence, we can write the vector $\mathbf{x}(k, m)$ as

$$\begin{aligned} \mathbf{x}(k, m) &= X(k, m)\boldsymbol{\rho}_X^*(k, m) + \mathbf{x}_{\mathrm{i}}(k, m) \\ &= \mathbf{x}_{\mathrm{d}}(k, m) + \mathbf{x}_{\mathrm{i}}(k, m), \end{aligned} \tag{3.9}$$

where

$$\mathbf{x}_{\mathrm{d}}(k, m) = X(k, m)\boldsymbol{\rho}_X^*(k, m) \tag{3.10}$$

is the desired signal vector,

$$\mathbf{x}_\mathrm{i}(k,m) = [X_\mathrm{i}(k,m)\ X_\mathrm{i}(k,m-1) \cdots X_\mathrm{i}(k,m-L+1)]^T$$

is the interference signal vector, and

$$\begin{aligned}\boldsymbol{\rho}_X(k,m) &= [\rho_X(k,m,0)\rho_X(k,m,1) \cdots \rho_X(k,m,L-1)]^T \\ &= [1\ \rho_X(k,m,1) \cdots \rho_X(k,m,L-1)]^T \\ &= \frac{E\left[X(k,m)\mathbf{x}^*(k,m)\right]}{E\left[|X(k,m)|^2\right]} \end{aligned} \tag{3.11}$$

is the (normalized) interframe correlation vector between $X(k,m)$ and $\mathbf{x}(k,m)$.
Substituting (3.9) into (3.2), we get

$$\begin{aligned} Z(k,m) &= \mathbf{h}^H(k,m)[X(k,m)\boldsymbol{\rho}_X^*(k,m) + \mathbf{x}_\mathrm{i}(k,m) + \mathbf{v}(k,m)] \\ &= X_\mathrm{fd}(k,m) + X_\mathrm{ri}(k,m) + V_\mathrm{rn}(k,m), \end{aligned} \tag{3.12}$$

where

$$X_\mathrm{fd}(k,m) = X(k,m)\mathbf{h}^H(k,m)\boldsymbol{\rho}_X^*(k,m) \tag{3.13}$$

is the filtered desired signal and

$$X_\mathrm{ri}(k,m) = \mathbf{h}^H(k,m)\mathbf{x}_\mathrm{i}(k,m) \tag{3.14}$$

is the residual interference. We observe that the estimate of the desired signal is the sum of three terms that are mutually uncorrelated. The first one is clearly the filtered desired signal while the two others are the filtered undesired signals (interference-plus-noise). Therefore, the variance of $Z(k,m)$ is

$$\begin{aligned} \phi_Z(k,m) &= \mathbf{h}^H(k,m)\boldsymbol{\Phi}_\mathbf{y}(k,m)\mathbf{h}(k,m) \\ &= \phi_{X_\mathrm{fd}}(k,m) + \phi_{X_\mathrm{ri}}(k,m) + \phi_{V_\mathrm{rn}}(k,m), \end{aligned} \tag{3.15}$$

where

$$\boldsymbol{\Phi}_\mathbf{y}(k,m) = E\left[\mathbf{y}(k,m)\mathbf{y}^H(k,m)\right] \tag{3.16}$$

is the correlation matrix of $\mathbf{y}(k,m)$,

$$\begin{aligned} \phi_{X_\mathrm{fd}}(k,m) &= \phi_X(k,m)\left|\mathbf{h}^H(k,m)\boldsymbol{\rho}_X^*(k,m)\right|^2 \\ &= \mathbf{h}^H(k,m)\boldsymbol{\Phi}_{\mathbf{x}_\mathrm{d}}(k,m)\mathbf{h}(k,m), \end{aligned} \tag{3.17}$$

$$\begin{aligned}\phi_{X_{\mathrm{ri}}}(k,m) &= \mathbf{h}^H(k,m)\mathbf{\Phi}_{\mathbf{x}_\mathrm{i}}(k,m)\mathbf{h}(k,m) \\ &= \mathbf{h}^H(k,m)\mathbf{\Phi}_{\mathbf{x}}(k,m)\mathbf{h}(k,m) - \phi_X(k,m)\left|\mathbf{h}^H(k,m)\boldsymbol{\rho}_X^*(k,m)\right|^2,\end{aligned} \tag{3.18}$$

$$\phi_{V_{\mathrm{rn}}}(k,m) = \mathbf{h}^H(k,m)\mathbf{\Phi}_{\mathbf{v}}(k,m)\mathbf{h}(k,m), \tag{3.19}$$

$$\mathbf{\Phi}_{\mathbf{x}_\mathrm{d}}(k,m) = \phi_X(k,m)\boldsymbol{\rho}_X^*(k,m)\boldsymbol{\rho}_X^T(k,m) \tag{3.20}$$

is the correlation matrix (whose rank is equal to 1) of $\mathbf{x}_\mathrm{d}(k,m)$ and

$$\mathbf{\Phi}_{\mathbf{a}}(k,m) = E\left[\mathbf{a}(k,m)\mathbf{a}^H(k,m)\right] \tag{3.21}$$

is the correlation matrix of $\mathbf{a}(k,m) \in \{\mathbf{x}(k,m), \mathbf{x}_\mathrm{i}(k,m), \mathbf{v}(k,m)\}$. In the rest, it is assumed that the rank of $\mathbf{\Phi}_{\mathbf{v}}(k,m)$ is L so that its inverse exists.

In some situations, it may also be useful to decompose the vector noise into two orthogonal components: one correlated and another uncorrelated with $V(k,m)$. We obtain

$$\mathbf{v}(k,m) = V(k,m)\boldsymbol{\rho}_V^*(k,m) + \mathbf{v}_\mathrm{u}(k,m), \tag{3.22}$$

where $\boldsymbol{\rho}_V(k,m)$ and $\mathbf{v}_\mathrm{u}(k,m)$ are defined in a similar way to $\boldsymbol{\rho}_X(k,m)$ and $\mathbf{x}_\mathrm{i}(k,m)$. Expression (3.12) becomes

$$\begin{aligned}Z(k,m) = {} & X(k,m)\mathbf{h}^H(k,m)\boldsymbol{\rho}_X^*(k,m) + V(k,m)\mathbf{h}^H(k,m)\boldsymbol{\rho}_V^*(k,m) \\ & + \mathbf{h}^H[\mathbf{x}_\mathrm{i}(k,m) + \mathbf{v}_\mathrm{u}(k,m)].\end{aligned} \tag{3.23}$$

Here again, $Z(k,m)$ is the sum of three mutually uncorrelated components. With this formulation, more constraints are possible on $\mathbf{h}(k,m)$.

3.2 Performance Measures

In this section, the performance measures which are tailored for the interframe correlation are defined.

3.2.1 Noise Reduction

The subband and fullband input SNRs were already defined in Chap. 2.

To quantify the level of noise remaining at the output of the FIR filter, we define the subband output SNR as[1]

[1] In this work, we consider the interference as part of the noise in the definitions of the performance measures.

$$\begin{aligned}\text{oSNR}\left[\mathbf{h}(k,m)\right] &= \frac{\phi_{X_{\text{fd}}}(k,m)}{\phi_{X_{\text{ri}}}(k,m)+\phi_{V_{\text{rn}}}(k,m)} \\ &= \frac{\phi_X(k,m)\left|\mathbf{h}^H(k,m)\boldsymbol{\rho}_X^*(k,m)\right|^2}{\mathbf{h}^H(k,m)\boldsymbol{\Phi}_{\text{in}}(k,m)\mathbf{h}(k,m)}, \quad k=0,1,\ldots,K-1,\end{aligned} \tag{3.24}$$

where

$$\boldsymbol{\Phi}_{\text{in}}(k,m) = \boldsymbol{\Phi}_{\mathbf{x}_{\text{i}}}(k,m) + \boldsymbol{\Phi}_{\mathbf{v}}(k,m) \tag{3.25}$$

is the interference-plus-noise correlation matrix. For the particular filter $\mathbf{h}(k,m) = \mathbf{i}_{L,1}$ (identity filter), where $\mathbf{i}_{L,1}$ is the first column of the identity matrix $\mathbf{I}_L$ (of size $L \times L$), we have

$$\text{oSNR}[\mathbf{i}_{L,1}(k,m)] = \text{iSNR}(k,m). \tag{3.26}$$

And for the particular case $L = 1$, we also have

$$\text{oSNR}[H_0(k,m)] = \text{iSNR}(k,m). \tag{3.27}$$

Hence, in the two previous scenarios, the subband SNR cannot be improved.

Now, let us define the quantity

$$\begin{aligned}\text{oSNR}_{\max}(k,m) &= \text{tr}[\boldsymbol{\Phi}_{\text{in}}^{-1}(k,m)\boldsymbol{\Phi}_{\mathbf{x}_{\text{d}}}(k,m)] \\ &= \phi_X(k,m)\boldsymbol{\rho}_X^T(k,m)\boldsymbol{\Phi}_{\text{in}}^{-1}(k,m)\boldsymbol{\rho}_X^*(k,m),\end{aligned} \tag{3.28}$$

where $\text{tr}[\cdot]$ denotes the trace of a square matrix. This quantity corresponds to the maximum eigenvalue, $\lambda_{\max}(k,m)$, of the matrix $\boldsymbol{\Phi}_{\text{in}}^{-1}(k,m)\boldsymbol{\Phi}_{\mathbf{x}_{\text{d}}}(k,m)$. It also corresponds to the maximum output SNR since the filter, $\mathbf{h}_{\max}(k,m)$, that maximizes $\text{oSNR}[\mathbf{h}(k,m)]$ [Eq. (3.24)] is the maximum eigenvector of $\boldsymbol{\Phi}_{\text{in}}^{-1}(k,m)\boldsymbol{\Phi}_{\mathbf{x}_{\text{d}}}(k,m)$ for which its corresponding eigenvalue is $\lambda_{\max}(k,m)$. As a result, we have

$$\text{oSNR}[\mathbf{h}(k,m)] \leq \text{oSNR}_{\max}(k,m) = \lambda_{\max}(k,m), \ \forall \mathbf{h}(k,m) \tag{3.29}$$

and

$$\text{oSNR}_{\max}(k,m) = \text{oSNR}[\mathbf{h}_{\max}(k,m)] \geq \text{oSNR}[\mathbf{i}_{L,1}(k,m)] = \text{iSNR}(k,m). \tag{3.30}$$

The maximum SNR filter is then

$$\mathbf{h}_{\max}(k,m) = \alpha(k,m)\boldsymbol{\Phi}_{\text{in}}^{-1}(k,m)\boldsymbol{\rho}_X^*(k,m), \tag{3.31}$$

where $\alpha(k,m) \neq 0$ is an arbitrary scaling factor. We will show in the next section that most of the optimal filters are equivalent to $\mathbf{h}_{\max}(k,m)$ up to that scaling factor.

We define the fullband output SNR at time-frame m as

$$\text{oSNR}[\mathbf{h}(:, m)] = \frac{\sum_{k=0}^{K-1} \phi_X(k, m) \left|\mathbf{h}^H(k, m)\boldsymbol{\rho}_X^*(k, m)\right|^2}{\sum_{k=0}^{K-1} \mathbf{h}^H(k, m)\boldsymbol{\Phi}_{\text{in}}(k, m)\mathbf{h}(k, m)} \tag{3.32}$$

and it can be verified that

$$\text{oSNR}[\mathbf{h}(:, m)] \leq \sum_{k=0}^{K-1} \text{oSNR}[\mathbf{h}(k, m)]. \tag{3.33}$$

As a result,

$$\text{oSNR}[\mathbf{h}(:, m)] \leq \sum_{k=0}^{K-1} \lambda_{\max}(k, m), \ \forall \mathbf{h}(k, m). \tag{3.34}$$

The fullband output SNR with the maximum SNR filter is

$$\text{oSNR}[\mathbf{h}_{\max}(:, m)] = \frac{\sum_{k=0}^{K-1} \frac{|\alpha(k, m)|^2 \lambda_{\max}^2(k, m)}{\phi_X(k, m)}}{\sum_{k=0}^{K-1} \frac{|\alpha(k, m)|^2 \lambda_{\max}(k, m)}{\phi_X(k, m)}}. \tag{3.35}$$

We see that the performance (in terms of SNR improvement) of the maximum SNR filter is quite dependent on the values of $\alpha(k, m)$. We can express (3.35) as

$$\text{oSNR}[\mathbf{h}_{\max}(:, m)] = \frac{\boldsymbol{\alpha}^H(m)\mathbf{D}_1(m)\boldsymbol{\alpha}(m)}{\boldsymbol{\alpha}^H(m)\mathbf{D}_2(m)\boldsymbol{\alpha}(m)}, \tag{3.36}$$

where

$$\boldsymbol{\alpha}(m) = [\alpha(0, m)\, \alpha(1, m) \cdots \alpha(K-1, m)]^T \tag{3.37}$$

is a vector of length K containing all the scaling factors and

$$\mathbf{D}_1(m) = \text{diag}\left[\frac{\lambda_{\max}^2(0, m)}{\phi_X(0, m)}, \frac{\lambda_{\max}^2(1, m)}{\phi_X(1, m)}, \ldots, \frac{\lambda_{\max}^2(K-1, m)}{\phi_X(K-1, m)}\right], \tag{3.38}$$

$$\mathbf{D}_2(m) = \text{diag}\left[\frac{\lambda_{\max}(0, m)}{\phi_X(0, m)}, \frac{\lambda_{\max}(1, m)}{\phi_X(1, m)}, \ldots, \frac{\lambda_{\max}(K-1, m)}{\phi_X(K-1, m)}\right], \tag{3.39}$$

are two diagonal matrices. Now, if we maximize (3.36) with respect to $\boldsymbol{\alpha}(m)$, we find that the solution, $\boldsymbol{\alpha}_{\max}(m)$, is the eigenvector corresponding to the maximum eigenvalue of the matrix $\mathbf{D}_2^{-1}(m)\mathbf{D}_1(m)$. Since this matrix is diagonal, its maximum eigenvalue is its largest diagonal element, i.e., $\max_k \ \lambda_{\max}(k, m)$. We deduce that

$$\text{oSNR}[\mathbf{h}(:, m)] \leq \max_k \ \lambda_{\max}(k, m), \quad \forall \mathbf{h}(k, m), \tag{3.40}$$

which is a much tighter bound than (3.34). This result is very interesting for two reasons. First, (3.40) shows how the fullband output SNR is upper bounded and this upper bound can never exceed the maximum subband output SNR. Second, if we compare this upper bound to the one given in Chap. 2, we see that the one given in (3.40) is larger. As a consequence, we can expect better output SNRs with the filter approach than with the gain approach.

The noise reduction factor quantifies the amount of noise that is rejected by the filter. The subband and fullband noise reduction factors are then

$$\begin{aligned}\xi_{\mathrm{nr}}[\mathbf{h}(k,m)] &= \frac{\phi_V(k,m)}{\phi_{X_{\mathrm{ri}}}(k,m)+\phi_{V_{\mathrm{rn}}}(k,m)}\\ &= \frac{\phi_V(k,m)}{\mathbf{h}^H(k,m)\mathbf{\Phi}_{\mathrm{in}}(k,m)\mathbf{h}(k,m)},\quad k=0,1,\ldots,K-1,\end{aligned} \tag{3.41}$$

$$\xi_{\mathrm{nr}}[\mathbf{h}(:,m)] = \frac{\sum_{k=0}^{K-1}\phi_V(k,m)}{\sum_{k=0}^{K-1}\mathbf{h}^H(k,m)\mathbf{\Phi}_{\mathrm{in}}(k,m)\mathbf{h}(k,m)}. \tag{3.42}$$

The noise reduction factors are expected to be lower bounded by 1 for optimal filters. So the more the noise is reduced, the higher are the values of the noise reduction factors.

3.2.2 Speech Distortion

The desired speech signal can be distorted by the filter. Therefore, we define the subband and fullband speech reduction factors as

$$\begin{aligned}\xi_{\mathrm{sr}}[\mathbf{h}(k,m)] &= \frac{\phi_X(k,m)}{\phi_{X_{\mathrm{fd}}}(k,m)}\\ &= \frac{1}{|\mathbf{h}^H(k,m)\boldsymbol{\rho}_X^*(k,m)|^2},\quad k=0,1,\ldots,K-1,\end{aligned} \tag{3.43}$$

$$\xi_{\mathrm{sr}}[\mathbf{h}(:,m)] = \frac{\sum_{k=0}^{K-1}\phi_X(k,m)}{\sum_{k=0}^{K-1}\phi_X(k,m)\left|\mathbf{h}^H(k,m)\boldsymbol{\rho}_X^*(k,m)\right|^2}. \tag{3.44}$$

An important observation is that the design of a filter that does not distort the desired signal requires the constraint

$$\mathbf{h}^H(k,m)\boldsymbol{\rho}_X^*(k,m) = 1,\quad \forall k,m. \tag{3.45}$$

Thus, the speech reduction factor is equal to 1 if there is no distortion and expected to be greater than 1 when distortion occurs.

It is easy to verify the fundamental relations:

$$\frac{\text{oSNR}[\mathbf{h}(k,m)]}{\text{iSNR}(k,m)} = \frac{\xi_{\text{nr}}[\mathbf{h}(k,m)]}{\xi_{\text{sr}}[\mathbf{h}(k,m)]},\ k = 0, 1, \ldots, K-1, \tag{3.46}$$

$$\frac{\text{oSNR}[\mathbf{h}(:,m)]}{\text{iSNR}(m)} = \frac{\xi_{\text{nr}}[\mathbf{h}(:,m)]}{\xi_{\text{sr}}[\mathbf{h}(:,m)]}. \tag{3.47}$$

Another useful performance measure is the speech distortion index defined as

$$\begin{aligned} \upsilon_{\text{sd}}[\mathbf{h}(k,m)] &= \frac{E\left\{|X_{\text{fd}}(k,m) - X(k,m)|^2\right\}}{\phi_X(k,m)} \\ &= \left|\mathbf{h}^H(k,m)\boldsymbol{\rho}_X^*(k,m) - 1\right|^2, \quad k = 0, 1, \ldots, K-1 \end{aligned} \tag{3.48}$$

in the subband case and as

$$\upsilon_{\text{sd}}[\mathbf{h}(:,m)] = \frac{\sum_{k=0}^{K-1} E\left\{|X_{\text{fd}}(k,m) - X(k,m)|^2\right\}}{\sum_{k=0}^{K-1} \phi_X(k,m)} \tag{3.49}$$

in the fullband case. The speech distortion index is always greater than or equal to 0 and should be upper bounded by 1 for optimal filters; so the higher is its value, the more the desired signal is distorted.

3.2.3 MSE Criterion

The error signal between the estimated and desired signals at the frequency-bin k and time-frame m is

$$\begin{aligned} \mathcal{E}(k,m) &= Z(k,m) - X(k,m) \\ &= \mathbf{h}^H(k,m)\mathbf{y}(k,m) - X(k,m). \end{aligned} \tag{3.50}$$

We can rewrite (3.50) as

$$\mathcal{E}(k,m) = \mathcal{E}_{\text{d}}(k,m) + \mathcal{E}_{\text{r}}(k,m), \tag{3.51}$$

where

$$\begin{aligned} \mathcal{E}_{\text{d}}(k,m) &= X_{\text{fd}}(k,m) - X(k,m) \\ &= \left[\mathbf{h}^H(k,m)\boldsymbol{\rho}_X^*(k,m) - 1\right] X(k,m) \end{aligned} \tag{3.52}$$

is the speech distortion due to the complex filter and

$$\begin{aligned} \mathcal{E}_{\text{r}}(k,m) &= X_{\text{ri}}(k,m) + V_{\text{rn}}(k,m) \\ &= \mathbf{h}^H(k,m)\mathbf{x}_{\text{i}}(k,m) + \mathbf{h}^H(k,m)\mathbf{v}(k,m) \end{aligned} \tag{3.53}$$

represents the residual interference-plus-noise.

Having defined the error signal, we can now write the subband MSE criterion:

$$\begin{aligned} J[\mathbf{h}(k,m)] &= E\left[|\mathcal{E}(k,m)|^2\right] \\ &= J_{\mathrm{d}}[\mathbf{h}(k,m)] + J_{\mathrm{r}}[\mathbf{h}(k,m)], \end{aligned} \tag{3.54}$$

where

$$\begin{aligned} J_{\mathrm{d}}\left[\mathbf{h}(k,m)\right] &= E\left[|\mathcal{E}_{\mathrm{d}}(k,m)|^2\right] \\ &= E\left[|X_{\mathrm{fd}}(k,m) - X(k,m)|^2\right] \\ &= \phi_X(k,m)\left|\mathbf{h}^H(k,m)\boldsymbol{\rho}_X^*(k,m) - 1\right|^2 \end{aligned} \tag{3.55}$$

and

$$\begin{aligned} J_{\mathrm{r}}[\mathbf{h}(k,m)] &= E\left[|\mathcal{E}_{\mathrm{r}}(k,m)|^2\right] \\ &= E\left[|X_{\mathrm{ri}}(k,m)|^2\right] + E\left[|V_{\mathrm{rn}}(k,m)|^2\right] \\ &= \phi_{X_{\mathrm{ri}}}(k,m) + \phi_{V_{\mathrm{rn}}}(k,m). \end{aligned} \tag{3.56}$$

For the two particular filters $\mathbf{h}(k,m) = \mathbf{i}_{L,1}$ and $\mathbf{h}(k,m) = \mathbf{0}_{L\times 1}$, we get

$$J[\mathbf{i}_{L,1}(k,m)] = J_{\mathrm{r}}[\mathbf{i}_{L,1}(k,m)] = \phi_V(k,m), \tag{3.57}$$

$$J[\mathbf{0}_{L\times 1}(k,m)] = J_{\mathrm{d}}[\mathbf{0}_{L\times 1}(k,m)] = \phi_X(k,m). \tag{3.58}$$

We then find that the subband NMSE with respect to $J[\mathbf{i}_{L,1}(k,m)]$ is

$$\begin{aligned} \widetilde{J}[\mathbf{h}(k,m)] &= \frac{J[\mathbf{h}(k,m)]}{J[\mathbf{i}_{L,1}(k,m)]} \\ &= \mathrm{iSNR}(k,m) \cdot \upsilon_{\mathrm{sd}}[\mathbf{h}(k,m)] + \frac{1}{\xi_{\mathrm{nr}}[\mathbf{h}(k,m)]}, \end{aligned} \tag{3.59}$$

where

$$\upsilon_{\mathrm{sd}}[\mathbf{h}(k,m)] = \frac{J_{\mathrm{d}}[\mathbf{h}(k,m)]}{J_{\mathrm{d}}[\mathbf{0}_{L\times 1}(k,m)]}, \tag{3.60}$$

$$\xi_{\mathrm{nr}}[\mathbf{h}(k,m)] = \frac{J_{\mathrm{r}}[\mathbf{i}_{L,1}(k,m)]}{J_{\mathrm{r}}[\mathbf{h}(k,m)]}, \tag{3.61}$$

and the subband NMSE with respect to $J[\mathbf{0}_{L\times 1}(k,m)]$ is

$$\begin{aligned} \overline{J}[\mathbf{h}(k,m)] &= \frac{J[\mathbf{h}(k,m)]}{J[\mathbf{0}_{L\times 1}(k,m)]} \\ &= \upsilon_{\mathrm{sd}}[\mathbf{h}(k,m)] + \frac{1}{\mathrm{oSNR}[\mathbf{h}(k,m)] \cdot \xi_{\mathrm{sr}}[\mathbf{h}(k,m)]}. \end{aligned} \tag{3.62}$$

We have

$$\widetilde{J}[\mathbf{h}(k,m)] = \text{iSNR}(k,m) \cdot \overline{J}[\mathbf{h}(k,m)]. \tag{3.63}$$

It is also easy to deduce that the fullband MSE and NMSEs at time-frame m are

$$\begin{aligned} J[\mathbf{h}(:,m)] &= \frac{1}{K}\sum_{k=0}^{K-1} J[\mathbf{h}(k,m)] \\ &= \frac{1}{K}\sum_{k=0}^{K-1} J_{\text{d}}[\mathbf{h}(k,m)] + \frac{1}{K}\sum_{k=0}^{K-1} J_{\text{r}}[\mathbf{h}(k,m)] \\ &= J_{\text{d}}[\mathbf{h}(:,m)] + J_{\text{r}}[\mathbf{h}(:,m)], \end{aligned} \tag{3.64}$$

$$\begin{aligned} \widetilde{J}[\mathbf{h}(:,m)] &= K\frac{J[\mathbf{h}(:,m)]}{\sum_{k=0}^{K-1}\phi_V(k,m)} \\ &= \text{iSNR}(m) \cdot \upsilon_{\text{sd}}[\mathbf{h}(:,m)] + \frac{1}{\xi_{\text{nr}}[\mathbf{h}(:,m)]}, \end{aligned} \tag{3.65}$$

$$\begin{aligned} \overline{J}[\mathbf{h}(:,m)] &= K\frac{J[\mathbf{h}(:,m)]}{\sum_{k=0}^{K-1}\phi_X(k,m)} \\ &= \upsilon_{\text{sd}}[\mathbf{h}(:,m)] + \frac{1}{\text{oSNR}[\mathbf{h}(:,m)] \cdot \xi_{\text{sr}}[\mathbf{h}(:,m)]}. \end{aligned} \tag{3.66}$$

It is clear that the objective of noise reduction in the STFT domain with the interframe filtering is to find optimal filters $\mathbf{h}(k,m)$ at each frequency-bin k and time-frame m that would either directly minimize $J[\mathbf{h}(k,m)]$ or minimize $J_{\text{d}}[\mathbf{h}(k,m)]$ or $J_{\text{r}}[\mathbf{h}(k,m)]$ subject to some constraint.

3.3 Optimal Filters

In this section, we are going to derive the most important filters that can help reduce the noise picked up by the microphone.

3.3.1 Wiener

The Wiener filter is easily derived by taking the gradient of the subband MSE, $J[\mathbf{h}(k,m)]$, with respect to $\mathbf{h}^H(k,m)$ and equating the result to zero:

$$\mathbf{h}_{\text{W}}(k,m) = \mathbf{\Phi}_{\mathbf{y}}^{-1}(k,m)\mathbf{\Phi}_{\mathbf{yx}}(k,m)\mathbf{i}_{L,1}, \tag{3.67}$$

where

$$\mathbf{\Phi}_{\mathbf{yx}}(k,m) = E\left[\mathbf{y}(k,m)\mathbf{x}^H(k,m)\right] \tag{3.68}$$

is the cross-correlation matrix between $\mathbf{y}(k,m)$ and $\mathbf{x}(k,m)$. But

$$\mathbf{\Phi}_{\mathbf{yx}}(k,m)\mathbf{i}_{L,1} = \phi_X(k,m)\boldsymbol{\rho}_X^*(k,m), \tag{3.69}$$

so that (3.67) becomes

$$\mathbf{h}_{\mathrm{W}}(k,m) = \phi_X(k,m)\mathbf{\Phi}_{\mathbf{y}}^{-1}(k,m)\boldsymbol{\rho}_X^*(k,m). \tag{3.70}$$

The Wiener filter can also be rewritten as

$$\begin{aligned}\mathbf{h}_{\mathrm{W}}(k,m) &= \mathbf{\Phi}_{\mathbf{y}}^{-1}(k,m)\mathbf{\Phi}_{\mathbf{x}}(k,m)\mathbf{i}_{L,1} \\ &= \left[\mathbf{I}_L - \mathbf{\Phi}_{\mathbf{y}}^{-1}(k,m)\mathbf{\Phi}_{\mathbf{v}}(k,m)\right]\mathbf{i}_{L,1}.\end{aligned} \tag{3.71}$$

We know that

$$\mathbf{\Phi}_{\mathbf{y}}(k,m) = \phi_X(k,m)\boldsymbol{\rho}_X^*(k,m)\boldsymbol{\rho}_X^T(k,m) + \mathbf{\Phi}_{\mathrm{in}}(k,m). \tag{3.72}$$

Determining the inverse of $\mathbf{\Phi}_{\mathbf{y}}(k,m)$ from (3.72) with the Woodbury's identity, we get

$$\mathbf{\Phi}_{\mathbf{y}}^{-1}(k,m) = \mathbf{\Phi}_{\mathrm{in}}^{-1}(k,m) - \frac{\mathbf{\Phi}_{\mathrm{in}}^{-1}(k,m)\boldsymbol{\rho}_X^*(k,m)\boldsymbol{\rho}_X^T(k,m)\mathbf{\Phi}_{\mathrm{in}}^{-1}(k,m)}{\phi_X^{-1}(k,m) + \boldsymbol{\rho}_X^T(k,m)\mathbf{\Phi}_{\mathrm{in}}^{-1}(k,m)\boldsymbol{\rho}_X^*(k,m)}. \tag{3.73}$$

Substituting this result into (3.70) leads to another interesting formulation of the Wiener filter:

$$\mathbf{h}_{\mathrm{W}}(k,m) = \frac{\phi_X(k,m)\mathbf{\Phi}_{\mathrm{in}}^{-1}(k,m)\boldsymbol{\rho}_X^*(k,m)}{1+\lambda_{\max}(k,m)}, \tag{3.74}$$

that we can rewrite as

$$\mathbf{h}_{\mathrm{W}}(k,m) = \frac{\mathbf{\Phi}_{\mathrm{in}}^{-1}(k,m)\mathbf{\Phi}_{\mathbf{y}}(k,m) - \mathbf{I}_L}{1 - L + \mathrm{tr}\left[\mathbf{\Phi}_{\mathrm{in}}^{-1}(k,m)\mathbf{\Phi}_{\mathbf{y}}(k,m)\right]}\mathbf{i}_{L,1}. \tag{3.75}$$

Using (3.74), we find that the subband output SNR is

$$\begin{aligned}\mathrm{oSNR}[\mathbf{h}_{\mathrm{W}}(k,m)] &= \lambda_{\max}(k,m) \\ &= \mathrm{tr}\left[\mathbf{\Phi}_{\mathrm{in}}^{-1}(k,m)\mathbf{\Phi}_{\mathbf{y}}(k,m)\right] - L\end{aligned} \tag{3.76}$$

and the subband speech distortion index is a clear function of this subband output SNR:

$$\upsilon_{\text{sd}}[\mathbf{h}_{\text{W}}(k,m)] = \frac{1}{\{1 + \text{oSNR}[\mathbf{h}_{\text{W}}(k,m)]\}^2}. \tag{3.77}$$

Interestingly, the higher is the value of $\text{oSNR}[\mathbf{h}_{\text{W}}(k,m)]$ (i.e., by increasing the number of interframes), the less the desired signal is distorted with the Wiener filter at frequency-bin k.

Clearly,

$$\text{oSNR}[\mathbf{h}_{\text{W}}(k,m)] \geq \text{iSNR}(k,m), \tag{3.78}$$

since the Wiener filter maximizes the subband output SNR. It is of great interest to observe that the two filters $\mathbf{h}_{\text{W}}(k,m)$ and $\mathbf{h}_{\max}(k,m)$ are equivalent up to a scaling factor. Indeed, taking

$$\alpha(k,m) = \frac{\phi_X(k,m)}{1 + \lambda_{\max}(k,m)} \tag{3.79}$$

in (3.31) (maximum SNR filter), we find (3.74) (Wiener filter).

With the Wiener filter, the subband noise reduction factor is

$$\begin{aligned} \xi_{\text{nr}}[\mathbf{h}_{\text{W}}(k,m)] &= \frac{[1 + \lambda_{\max}(k,m)]^2}{\text{iSNR}(k,m) \cdot \lambda_{\max}(k,m)} \\ &\geq \left[1 + \frac{1}{\lambda_{\max}(k,m)}\right]^2. \end{aligned} \tag{3.80}$$

Using (3.77) and (3.80) in (3.59), we find the minimum NMSE (MNMSE):

$$\widetilde{J}[\mathbf{h}_{\text{W}}(k,m)] = \frac{\text{iSNR}(k,m)}{1 + \lambda_{\max}(k,m)}. \tag{3.81}$$

It can be verified that the fullband output SNR with the Wiener filter at time-frame m is

$$\text{oSNR}[\mathbf{h}_{\text{W}}(:,m)] = \frac{\sum_{k=0}^{K-1} \frac{\lambda_{\max}^2(k,m)}{[1 + \lambda_{\max}(k,m)]^2} \phi_X(k,m)}{\sum_{k=0}^{K-1} \frac{\lambda_{\max}(k,m)}{[1 + \lambda_{\max}(k,m)]^2} \phi_X(k,m)}. \tag{3.82}$$

Property 3.1 *With the STFT-domain Wiener filter,* $\mathbf{h}_W(k,m)$, *the fullband output SNR is always greater than or equal to the fullband input SNR, i.e.,* $\text{oSNR}[\mathbf{h}_W(:,m)] \geq \text{iSNR}(m)$.

Proof See Sect. 3.3.4.

3.3.2 Minimum Variance Distortionless Response (MVDR)

The celebrated minimum variance distortionless response (MVDR) filter proposed by Capon [7, 8] is usually derived in a context where we have at least two sensors (or microphones) available. Remarkably, we can derive a very useful and practical MVDR when the interframe correlation is exploited by minimizing the MSE of the residual interference-plus-noise, $J_r[\mathbf{h}(k, m)]$, with the constraint that the desired signal is not distorted. Mathematically, this is equivalent to

$$\min_{\mathbf{h}(k,m)} \mathbf{h}^H(k, m)\mathbf{\Phi}_{\text{in}}(k, m)\mathbf{h}(k, m) \quad \text{subject to} \quad \mathbf{h}^H(k, m)\boldsymbol{\rho}_X^*(k, m) = 1, \tag{3.83}$$

for which the solution is

$$\begin{aligned}\mathbf{h}_{\text{MVDR}}(k, m) &= \frac{\phi_X(k, m)\mathbf{\Phi}_{\text{in}}^{-1}(k, m)\boldsymbol{\rho}_X^*(k, m)}{\lambda_{\max}(k, m)} \\ &= \frac{\mathbf{\Phi}_{\text{in}}^{-1}(k, m)\mathbf{\Phi}_{\mathbf{y}}(k, m) - \mathbf{I}_L}{\text{tr}\left[\mathbf{\Phi}_{\text{in}}^{-1}(k, m)\mathbf{\Phi}_{\mathbf{y}}(k, m)\right] - L}\mathbf{i}_{L,1}.\end{aligned} \tag{3.84}$$

Alternatively, we can express the MVDR as

$$\mathbf{h}_{\text{MVDR}}(k, m) = \frac{\mathbf{\Phi}_{\mathbf{y}}^{-1}(k, m)\boldsymbol{\rho}_X^*(k, m)}{\boldsymbol{\rho}_X^T(k, m)\mathbf{\Phi}_{\mathbf{y}}^{-1}(k, m)\boldsymbol{\rho}_X^*(k, m)}. \tag{3.85}$$

The Wiener and MVDR filters are simply related as follows

$$\mathbf{h}_{\text{W}}(k, m) = C(k, m)\mathbf{h}_{\text{MVDR}}(k, m), \tag{3.86}$$

where

$$C(k, m) = \frac{\lambda_{\max}(k, m)}{1 + \lambda_{\max}(k, m)}. \tag{3.87}$$

Here again the two filters $\mathbf{h}_{\text{W}}(k, m)$ and $\mathbf{h}_{\text{MVDR}}(k, m)$ are equivalent up to a scaling factor. From a subband point of view, this scaling is not significant but from a fullband point of view it can be important since speech signals are broadband in nature. Indeed, it can easily be verified that this scaling factor affects the fullband output SNRs and the speech distortion indices. While the subband output SNRs of the Wiener and MVDR filters are the same, the fullband output SNRs are not because of the scaling factor.

It is clear that we always have

$$\text{oSNR}[\mathbf{h}_{\text{MVDR}}(k, m)] = \text{oSNR}[\mathbf{h}_{\text{W}}(k, m)], \tag{3.88}$$

$$\upsilon_{\text{sd}}[\mathbf{h}_{\text{MVDR}}(k, m)] = 0, \tag{3.89}$$

$$\xi_{\text{sr}}[\mathbf{h}_{\text{MVDR}}(k, m)] = 1, \tag{3.90}$$

$$\xi_{\rm nr}[\mathbf{h}_{\rm MVDR}(k,m)] = \frac{\lambda_{\max}(k,m)}{\mathrm{iSNR}(k,m)} \le \xi_{\rm nr}[\mathbf{h}_{\rm W}(k,m)], \quad (3.91)$$

and

$$1 \ge \widetilde{J}[\mathbf{h}_{\rm MVDR}(k,m)] = \frac{\mathrm{iSNR}(k,m)}{\lambda_{\max}(k,m)} \ge \widetilde{J}[\mathbf{h}_{\rm W}(k,m)]. \quad (3.92)$$

The fullband output SNR with the MVDR filter is

$$\mathrm{oSNR}[\mathbf{h}_{\rm MVDR}(:,m)] = \frac{\sum_{k=0}^{K-1} \phi_X(k,m)}{\sum_{k=0}^{K-1} \frac{1}{\lambda_{\max}(k,m)} \phi_X(k,m)}. \quad (3.93)$$

Property 3.2 *With the STFT-domain MVDR filter,* $\mathbf{h}_{\rm MVDR}(k,m)$*, the fullband output SNR is always greater than or equal to the fullband input SNR, i.e.,* $\mathrm{oSNR}[\mathbf{h}_{\rm MVDR}(:,m)] \ge \mathrm{iSNR}(m)$.

Proof Indeed, we know that

$$\lambda_{\max}(k,m) \ge \mathrm{iSNR}(k,m) = \frac{\phi_X(k,m)}{\phi_V(k,m)}. \quad (3.94)$$

Hence,

$$\frac{\phi_X(k,m)}{\lambda_{\max}(k,m)} \le \phi_V(k,m). \quad (3.95)$$

Summing on k both sides of the previous expression and taking the inverse, we get

$$\frac{1}{\sum_{k=0}^{K-1} \frac{\phi_X(k,m)}{\lambda_{\max}(k,m)}} \ge \frac{1}{\sum_{k=0}^{K-1} \phi_V(k,m)} \quad (3.96)$$

and multiplying both sides of the previous equation by $\sum_{k=0}^{K-1} \phi_X(k,m)$, we finally obtain

$$\mathrm{oSNR}[\mathbf{h}_{\rm MVDR}(:,m)] \ge \mathrm{iSNR}(m), \quad (3.97)$$

which ends the proof.

3.3.3 Temporal Prediction

Assume that we can find a temporal prediction filter $\mathbf{h}'(k,m)$ of length L in such a way that

$$\mathbf{x}(k,m) \approx X(k,m)\mathbf{h}'(k,m). \quad (3.98)$$

The estimate of $X(k,m)$ becomes

$$Z(k,m) \approx X(k,m)\mathbf{h}^H(k,m)\mathbf{h}'(k,m) + \mathbf{h}^H(k,m)\mathbf{v}(k,m). \quad (3.99)$$

We can then derive a distortionless filter for speech enhancement as follows:

$$\min_{\mathbf{h}(k,m)} \phi_Z(k,m) \quad \text{subject to} \quad \mathbf{h}^H(k,m)\mathbf{h}'(k,m) = 1. \quad (3.100)$$

We easily deduce the solution

$$\mathbf{h}_{\mathrm{P}}(k,m) = \frac{\mathbf{\Phi}_{\mathbf{y}}^{-1}(k,m)\mathbf{h}'(k,m)}{\mathbf{h}'^H(k,m)\mathbf{\Phi}_{\mathbf{y}}^{-1}(k,m)\mathbf{h}'(k,m)}. \quad (3.101)$$

Now, we can find the optimal $\mathbf{h}'(k,m)$ in the Wiener sense. For that, we need to define the error signal vector

$$\mathbf{e}_{\mathrm{P}}(k,m) = \mathbf{x}(k,m) - X(k,m)\mathbf{h}'(k,m) \quad (3.102)$$

and form the MSE

$$J[\mathbf{h}'(k,m)] = E[\mathbf{e}_{\mathrm{P}}^H(k,m)\mathbf{e}_{\mathrm{P}}(k,m)]. \quad (3.103)$$

By minimizing $J[\mathbf{h}'(k,m)]$ with respect to $\mathbf{h}'(k,m)$, we easily find the optimal filter

$$\mathbf{h}'_{\mathrm{o}}(k,m) = \boldsymbol{\rho}_X^*(k,m). \quad (3.104)$$

It is interesting to observe that the error signal vector with the optimal filter, $\mathbf{h}'_{\mathrm{o}}(k,m)$, corresponds to the interference signal vector, i.e.,

$$\begin{aligned} \mathbf{e}_{\mathrm{P,o}}(k,m) &= \mathbf{x}(k,m) - X(k,m)\boldsymbol{\rho}_X^*(k,m) \\ &= \mathbf{x}_{\mathrm{i}}(k,m). \end{aligned} \quad (3.105)$$

This result is obviously expected because of the orthogonality principle.

Substituting (3.104) into (3.101), we find that

$$\mathbf{h}_{\mathrm{P}}(k,m) = \frac{\mathbf{\Phi}_{\mathbf{y}}^{-1}(k,m)\boldsymbol{\rho}_X^*(k,m)}{\boldsymbol{\rho}_X^T(k,m)\mathbf{\Phi}_{\mathbf{y}}^{-1}(k,m)\boldsymbol{\rho}_X^*(k,m)}. \quad (3.106)$$

Clearly, the two filters $\mathbf{h}_{\mathrm{MVDR}}(k,m)$ and $\mathbf{h}_{\mathrm{P}}(k,m)$ are identical. Therefore, the prediction approach with (3.100) can be seen as another way to derive the MVDR. This approach is also an intuitive manner to justify the decomposition given in (3.9).

It is possible to find another temporal prediction filter by optimizing the criteria

$$\min_{\mathbf{h}(k,m)} \mathbf{h}^H(k,m)\mathbf{\Phi}_{\mathbf{v}}(k,m)\mathbf{h}(k,m) \quad \text{subject to} \quad \mathbf{h}^H(k,m)\mathbf{h}'(k,m) = 1. \tag{3.107}$$

This leads to

$$\mathbf{h}_{\mathrm{P},2}(k,m) = \frac{\mathbf{\Phi}_{\mathbf{v}}^{-1}(k,m)\boldsymbol{\rho}_X^*(k,m)}{\boldsymbol{\rho}_X^T(k,m)\mathbf{\Phi}_{\mathbf{v}}^{-1}(k,m)\boldsymbol{\rho}_X^*(k,m)} \tag{3.108}$$

and, this time, this second temporal prediction filter is different from the MVDR filter.

The filter $\mathbf{h}_{\mathrm{P},2}(k,m)$ is suboptimal. As a result, the quality of the enhanced signal obtained with this filter is not expected to be better than the quality of the enhanced signal obtained with the MVDR filter.

3.3.4 Tradeoff

In the tradeoff approach, we try to compromise between noise reduction and speech distortion. Here, we minimize the speech distortion index with the constraint that the noise reduction factor is equal to a positive value that is greater than 1. Mathematically, this is equivalent to

$$\min_{\mathbf{h}(k,m)} J_{\mathrm{d}}[\mathbf{h}(k,m)] \quad \text{subject to} \quad J_{\mathrm{r}}[\mathbf{h}(k,m)] = \beta\phi_V(k,m), \tag{3.109}$$

where $0 < \beta < 1$ to insure that we get some noise reduction. By using a Lagrange multiplier, $\mu > 0$, to adjoin the constraint to the cost function, we easily deduce the tradeoff filter:

$$\begin{aligned} \mathbf{h}_{\mathrm{T},\mu}(k,m) &= \phi_X(k,m)\left[\phi_X(k,m)\boldsymbol{\rho}_X^*(k,m)\boldsymbol{\rho}_X^T(k,m) + \mu\mathbf{\Phi}_{\mathrm{in}}(k,m)\right]^{-1} \\ &\quad \times \boldsymbol{\rho}_X^*(k,m) \\ &= \frac{\phi_X(k,m)\mathbf{\Phi}_{\mathrm{in}}^{-1}(k,m)\boldsymbol{\rho}_X^*(k,m)}{\mu + \lambda_{\max}(k,m)}, \end{aligned} \tag{3.110}$$

where the Lagrange multiplier, μ, satisfies $J_{\mathrm{r}}[\mathbf{h}_{\mathrm{T},\mu}(k,m)] = \beta\phi_V(k,m)$. However, in practice it is not easy to determine the optimal μ. Therefore, when this parameter is chosen in an ad-hoc way, we can see that for

- $\mu = 1$, $\mathbf{h}_{\mathrm{T},1}(k,m) = \mathbf{h}_{\mathrm{W}}(k,m)$, which is the Wiener filter;
- $\mu = 0$, $\mathbf{h}_{\mathrm{T},0}(k,m) = \mathbf{h}_{\mathrm{MVDR}}(k,m)$, which is the MVDR filter;
- $\mu > 1$, results in a filter with low residual noise at the expense of high speech distortion;
- $\mu < 1$, results in a filter with high residual noise and low speech distortion.

Again, we observe here as well that the tradeoff and Wiener filters are equivalent up to a scaling factor. As a result, the subband output SNR with the tradeoff filter is obviously the same as the subband output SNR with the Wiener filter, i.e.,

$$\text{oSNR}[\mathbf{h}_{\text{T},\mu}(k,m)] = \lambda_{\max}(k,m) \tag{3.111}$$

and does not depend on μ. However, the subband speech distortion index is now both a function of the variable μ and the subband output SNR:

$$\upsilon_{\text{sd}}[\mathbf{h}_{\text{T},\mu}(k,m)] = \frac{\mu^2}{[\mu + \lambda_{\max}(k,m)]^2}. \tag{3.112}$$

From (3.112), we observe how μ can affect the desired signal.

Since the Wiener and MVDR filters are particular cases of the tradeoff filter, it is then useful to study the fullband output SNR and the fullband speech distortion index of the tradeoff filter, which both depend on the variable μ.

Using (3.110) in (3.32), we find that the fullband output SNR is

$$\text{oSNR}[\mathbf{h}_{\text{T},\mu}(:,m)] = \frac{\sum_{k=0}^{K-1} \dfrac{\phi_X(k,m)\lambda_{\max}^2(k,m)}{[\mu + \lambda_{\max}(k,m)]^2}}{\sum_{k=0}^{K-1} \dfrac{\phi_X(k,m)\lambda_{\max}(k,m)}{[\mu + \lambda_{\max}(k,m)]^2}}. \tag{3.113}$$

We propose the following.

Property 3.3 *The fullband output SNR of the STFT-domain tradeoff filter is an increasing function of the parameter* μ.

Proof Indeed, using the proof given in [9] by simply replacing integrals by sums, we find that

$$\frac{d\text{oSNR}[\mathbf{h}_{\text{T},\mu}(:,m)]}{d\mu} \geq 0, \tag{3.114}$$

proving that the fullband output SNR is increasing when μ is increasing.

From Property 3.3, we deduce that the MVDR filter gives the smallest fullband output SNR, which is

$$\text{oSNR}[\mathbf{h}_{\text{T},0}(:,m)] = \frac{\sum_{k=0}^{K-1} \phi_X(k,m)}{\sum_{k=0}^{K-1} \dfrac{\phi_X(k,m)}{\lambda_{\max}(k,m)}}. \tag{3.115}$$

We give another interesting property.

Property 3.4 *We have*

$$\lim_{\mu\to\infty} \text{oSNR}\left[\mathbf{h}_{\text{T},\mu}(:,m)\right] = \frac{\sum_{k=0}^{K-1} \phi_X(k,m)\lambda_{\max}^2(k,m)}{\sum_{k=0}^{K-1} \phi_X(k,m)\lambda_{\max}(k,m)} \leq \max_k \lambda_{\max}(k,m). \tag{3.116}$$

Proof Easy to show from (3.113).

While the fullband output SNR is upper bounded, it is easy to show that the fullband noise reduction factor and fullband speech reduction factor are not. So when μ goes to infinity so are $\xi_{nr}[\mathbf{h}_{T,\mu}(:,m)]$ and $\xi_{sr}[\mathbf{h}_{T,\mu}(:,m)]$.

The fullband speech distortion index is

$$\upsilon_{sd}[\mathbf{h}_{T,\mu}(:,m)] = \frac{\sum_{k=0}^{K-1} \frac{\phi_X(k,m)\mu^2}{[\mu + \lambda_{max}(k,m)]^2}}{\sum_{k=0}^{K-1} \phi_X(k,m)}. \tag{3.117}$$

Property 3.5 *The fullband speech distortion index of the STFT-domain tradeoff filter is an increasing function of the parameter μ.*

Proof It is straightforward to verify that

$$\frac{d\upsilon_{sd}[\mathbf{h}_{T,\mu}(:,m)]}{d\mu} \geq 0, \tag{3.118}$$

which ends the proof.

It is clear that

$$0 \leq \upsilon_{sd}\ [\mathbf{h}_{T,\mu}(:,m)] \leq 1, \quad \forall\ \mu \geq 0. \tag{3.119}$$

Therefore, as μ increases, the fullband output SNR increases at the price of more distortion to the desired signal.

Property 3.6 *With the STFT-domain tradeoff filter, $\mathbf{h}_{T,\mu}(k,m)$, the fullband output SNR is always greater than or equal to the fullband input SNR, i.e.,* $\text{oSNR}[\mathbf{h}_{T,\mu}(:,m)] \geq \text{iSNR}(m), \forall\mu \geq 0$.

Proof Since

$$\frac{d\text{oSNR}[\mathbf{h}_{T,\mu}(:,m)]}{d\mu} \geq 0, \tag{3.120}$$

and he fullband output SNR of the MVDR is always greater than or equal to the fullband input SNR, we deduce that

$$\text{oSNR}[\mathbf{h}_{T,\mu}(:,m)] \geq \text{iSNR}(m), \quad \forall\mu \geq 0, \tag{3.121}$$

which completes the proof.

To end this section, let us mention that the decision-directed method [10], which is a reliable estimator of the subband input SNR, will fit very well with the proposed algorithms since this estimator implicitly assumes that the successive frames are correlated.

3.3.5 Linearly Constrained Minimum Variance (LCMV)

We can derive a linearly constrained minimum variance (LCMV) filter [11, 12], which can handle more than one linear constraint, by exploiting the decomposition of the noise signal (see Sect. 3.1):

$$\mathbf{v}(k,m) = V(k,m)\boldsymbol{\rho}_V^*(k,m) + \mathbf{v}_\mathrm{u}(k,m). \tag{3.122}$$

Our problem this time is the following. We wish to perfectly recover our desired signal, $X(k,m)$, and completely remove the correlated components, $V(k,m)$ $\boldsymbol{\rho}_V^*(k,m)$. Thus, the two constraints can be put together in a matrix form as

$$\mathbf{C}^H(k,m)\mathbf{h}(k,m) = \begin{bmatrix} 1 \\ 0 \end{bmatrix}, \tag{3.123}$$

where

$$\mathbf{C}(k,m) = [\boldsymbol{\rho}_X^*(k,m)\boldsymbol{\rho}_V^*(k,m)] \tag{3.124}$$

is our constraint matrix of size $L \times 2$. Then, our optimal filter is obtained by minimizing the energy at the filter output, with the constraints that the correlated noise components are cancelled and the desired speech is preserved, i.e.,

$$\mathbf{h}_\mathrm{LCMV}(k,m) = \arg\min_{\mathbf{h}(k,m)} \mathbf{h}^H(k,m)\mathbf{\Phi}_\mathbf{y}(k,m)\mathbf{h}(k,m)$$
$$\text{subject to} \quad \mathbf{C}^H(k,m)\mathbf{h}(k,m) = \begin{bmatrix} 1 \\ 0 \end{bmatrix}. \tag{3.125}$$

The solution to (3.125) is given by

$$\mathbf{h}_\mathrm{LCMV}(k,m) = \mathbf{\Phi}_\mathbf{y}^{-1}(k,m)\mathbf{C}(k,m)[\mathbf{C}^H(k,m)\mathbf{\Phi}_\mathbf{y}^{-1}(k,m)\mathbf{C}(k,m)]^{-1}\begin{bmatrix} 1 \\ 0 \end{bmatrix}. \tag{3.126}$$

By developing (3.126), it can easily be shown that the LCMV can be written as a function of the MVDR, i.e.,

$$\mathbf{h}_\mathrm{LCMV}(k,m) = \frac{1}{1-|\varpi(k,m)|^2}\mathbf{h}_\mathrm{MVDR}(k,m) - \frac{|\varpi(k,m)|^2}{1-|\varpi(k,m)|^2}\mathbf{t}(k,m), \tag{3.127}$$

where

$$|\varpi(k,m)|^2 = \frac{\left|\boldsymbol{\rho}_X^T(k,m)\mathbf{\Phi}_\mathbf{y}^{-1}(k,m)\boldsymbol{\rho}_V^*(k,m)\right|^2}{\left[\boldsymbol{\rho}_X^T(k,m)\mathbf{\Phi}_\mathbf{y}^{-1}(k,m)\boldsymbol{\rho}_X^*(k,m)\right]\left[\boldsymbol{\rho}_V^T(k,m)\mathbf{\Phi}_\mathbf{y}^{-1}(k,m)\boldsymbol{\rho}_V^*(k,m)\right]}, \tag{3.128}$$

with $0 \leq |\varpi(k,m)|^2 \leq 1$, $\mathbf{h}_{\mathrm{MVDR}}(k,m)$ is defined in (3.85), and

$$\mathbf{t}(k,m) = \frac{\mathbf{\Phi}_{\mathbf{y}}^{-1}(k,m)\boldsymbol{\rho}_V^*(k,m)}{\boldsymbol{\rho}_X^T(k,m)\mathbf{\Phi}_{\mathbf{y}}^{-1}(k,m)\boldsymbol{\rho}_V^*(k,m)}. \tag{3.129}$$

We observe from (3.127) that when $|\varpi(k,m)|^2$ approaches 0, the LCMV filter tends to the MVDR filter; however, when $|\varpi(k,m)|^2$ approaches 1, we have no solution since we have conflicting requirements.

Obviously, we always have

$$\mathrm{oSNR}[\mathbf{h}_{\mathrm{LCMV}}(k,m)] \leq \mathrm{oSNR}[\mathbf{h}_{\mathrm{MVDR}}(k,m)], \tag{3.130}$$

$$\upsilon_{\mathrm{sd}}[\mathbf{h}_{\mathrm{LCMV}}(k,m)] = 0, \tag{3.131}$$

$$\xi_{\mathrm{sr}}[\mathbf{h}_{\mathrm{LCMV}}(k,m)] = 1, \tag{3.132}$$

and

$$\xi_{\mathrm{nr}}[\mathbf{h}_{\mathrm{LCMV}}(k,m)] \leq \xi_{\mathrm{nr}}[\mathbf{h}_{\mathrm{MVDR}}(k,m)] \leq \xi_{\mathrm{nr}}[\mathbf{h}_{\mathrm{W}}(k,m)]. \tag{3.133}$$

The LCMV filter is able to remove all the correlated noise but at the price that its overall noise reduction is lower than that of the MVDR filter.

References

1. J. Benesty, Y. Huang, *A Perspective on Single-Channel Frequency-Domain Speech Enhancement* (Morgan & Claypool Publishers, San Raphael, 2011)
2. J. Benesty, Y. Huang, A single-channel noise reduction MVDR filter, in *Proceedings of the IEEE ICASSP* (2011) pp. 273–276
3. J. Benesty, J. Chen, Y. Huang, I. Cohen, *Noise Reduction in Speech Processing* (Springer, Berlin, 2009)
4. J. Benesty, J. Chen, Y. Huang, On noise reduction in the Karhunen-Loève expansion domain, in *Proceedings of the IEEE ICASSP*, pp. 25–28 (2009)
5. J. Chen, J. Benesty, Y. Huang, Study of the noise-reduction problem in the Karhunen-Loève expansion domain. IEEE Trans. Audio Speech Lang. Process. **17**, 787–802 (2009)
6. I. Cohen, Relaxed statistical model for speech enhancement and a priori SNR estimation . IEEE Trans. Speech Audio Process. **13**, 870–881 (2005)
7. J. Capon, High resolution frequency-wavenumber spectrum analysis. Proc. IEEE **57**, 1408–1418 (1969)
8. R.T. Lacoss, Data adaptive spectral analysis methods. Geophysics **36**, 661–675 (1971)
9. M. Souden, J. Benesty, S. Affes, On the global output SNR of the parameterized frequency-domain multichannel noise reduction Wiener filter. IEEE Signal Process. Lett. **17**, 425–428 (2010)

10. Y. Ephraim, D. Malah, Speech enhancement using a minimum mean-square error short-time spectral amplitude estimator. IEEE Trans. Acoust. Speech Signal Process. **ASSP-32**, 1109–1121 (1984)
11. M. Er, A. Cantoni, Derivative constraints for broad-band element space antenna array processors. IEEE Trans. Acoust. Speech Signal Process. **31**, 1378–1393 (1983)
12. O. Frost, An algorithm for linearly constrained adaptive array processing. Proc. IEEE **60**, 926–935 (1972)

Chapter 4
Multichannel Speech Enhancement with Gains

In the previous two chapters, we exploited the temporal information from a single microphone signal to derive different techniques for speech enhancement in the STFT domain. In this chapter and the next one, we will exploit both the temporal and spatial information available from signals picked up by a determined number of microphones at different positions in the acoustics space in order to mitigate the noise effect. The focus of this chapter is on the derivation of optimal gains for noise reduction with a microphone array.

4.1 Signal Model

We consider the conventional signal model in which a microphone array with N sensors captures a convolved source signal in some noise field. The received signals are expressed as [1, 2]

$$\begin{aligned} y_n(t) &= g_n(t) * s(t) + v_n(t) \\ &= x_n(t) + v_n(t), n = 1, 2, \ldots, N, \end{aligned} \tag{4.1}$$

where $g_n(t)$ is the acoustic impulse response from the unknown speech source, $s(t)$, location to the nth microphone, $*$ stands for linear convolution, and $v_n(t)$ is the additive noise at microphone n. We assume that the impulse responses are time invariant. We also assume that the signals $x_n(t) = g_n(t) * s(t)$ and $v_n(t)$ are uncorrelated, zero mean, real, and broadband. By definition, $x_n(t)$ is coherent across the array. The noise signals, $v_n(t)$, are typically only partially coherent across the array.

In this work, our desired signal is designated by the clean (but convolved) speech signal received at microphone 1, namely $x_1(t)$. Obviously, any signal $x_n(t)$ could be taken as the reference. Our problem then may be stated as follows [3]: given N mixtures of two uncorrelated signals $x_n(t)$ and $v_n(t)$, our aim is to preserve $x_1(t)$ while minimizing the contribution of the noise terms, $v_n(t)$, at the array output.

J. Benesty et al., *Speech Enhancement in the STFT Domain*,
SpringerBriefs in Electrical and Computer Engineering,
DOI: 10.1007/978-3-642-23250-3_4, © The Author(s) 2012

Expression (4.1) can be rewritten in the STFT domain as

$$\begin{aligned} Y_n(k,m) &= G_n(k)S(k,m) + V_n(k,m) \\ &= X_n(k,m) + V_n(k,m),\ n = 1, 2, \ldots, N, \end{aligned} \tag{4.2}$$

where $Y_n(k,m),\ G_n(k),\ S(k,m),\ X_n(k,m) = G_n(k)S(k,m)$, and $V_n(k,m)$ are the STFT-domain representations of $y_n(t),\ g_n(t),\ s(t),\ x_n(t)$, and $v_n(t)$, respectively.

It is more convenient to write the N STFT-domain microphone signals in a vector notation:

$$\begin{aligned} \overleftarrow{\mathbf{y}}(k,m) &= \overleftarrow{\mathbf{g}}(k)S(k,m) + \overleftarrow{\mathbf{v}}(k,m) \\ &= \overleftarrow{\mathbf{x}}(k,m) + \overleftarrow{\mathbf{v}}(k,m) \\ &= \overleftarrow{\mathbf{d}}(k)X_1(k,m) + \overleftarrow{\mathbf{v}}(k,m), \end{aligned} \tag{4.3}$$

where

$$\begin{aligned} \overleftarrow{\mathbf{y}}(k,m) &= \left[\, Y_1(k,m)\ Y_2(k,m)\ \cdots\ Y_N(k,m)\,\right]^T, \\ \overleftarrow{\mathbf{x}}(k,m) &= \left[\, X_1(k,m)\ X_2(k,m)\ \cdots\ X_N(k,m)\,\right]^T \\ &= S(k,m)\overleftarrow{\mathbf{g}}(k), \\ \overleftarrow{\mathbf{g}}(k) &= \left[\, G_1(k)\ G_2(k)\ \cdots\ G_N(k)\,\right]^T, \\ \overleftarrow{\mathbf{v}}(k,m) &= \left[\, V_1(k,m)\ V_2(k,m)\ \cdots\ V_N(k,m)\,\right]^T, \end{aligned}$$

and

$$\begin{aligned} \overleftarrow{\mathbf{d}}(k) &= \left[\, 1\ \frac{G_2(k)}{G_1(k)}\ \cdots\ \frac{G_N(k)}{G_1(k)}\,\right]^T \\ &= \frac{\overleftarrow{\mathbf{g}}(k)}{G_1(k)}. \end{aligned} \tag{4.4}$$

Let us note that we assume that $G_1(k) \neq 0$. Expression (4.3) depends explicitly on the desired signal, $X_1(k,m)$; as a result, (4.3) is the STFT-domain signal model for noise reduction. The vector $\overleftarrow{\mathbf{d}}(k)$ is obviously the STFT-domain steering vector for noise reduction [3] since the acoustic impulse responses ratios from the broadband source to the aperture convey information about the position of the source.

There is another interesting way to write (4.3). First, it is easy to see that

$$X_n(k,m) = \rho^*_{X_1X_n}(k,m)X_1(k,m),\ n = 1, 2, \ldots, N, \tag{4.5}$$

where

$$\begin{aligned} \rho_{X_1X_n}(k,m) &= \frac{E\left[X_1(k,m)X_n^*(k,m)\right]}{E\left[|X_1(k,m)|^2\right]} \\ &= \frac{G_n^*(k)}{G_1^*(k)},\ n = 1, 2, \ldots, N \end{aligned} \tag{4.6}$$

is the partially normalized [with respect to $X_1(k, m)$] correlation coefficient between $X_1(k, m)$ and $X_n(k, m)$. Using (4.5), we can rewrite (4.3) as

$$\begin{aligned}\overleftarrow{\mathbf{y}}(k, m) &= \boldsymbol{\rho}^*_{X_1\overleftarrow{\mathbf{x}}}(k, m)X_1(k, m) + \overleftarrow{\mathbf{v}}(k, m)\\ &= \overleftarrow{\mathbf{x}}(k, m) + \overleftarrow{\mathbf{v}}(k, m),\end{aligned} \tag{4.7}$$

where

$$\overleftarrow{\mathbf{x}}(k, m) = \boldsymbol{\rho}^*_{X_1\overleftarrow{\mathbf{x}}}(k, m)X_1(k, m) \tag{4.8}$$

is the speech signal vector and

$$\begin{aligned}\boldsymbol{\rho}_{X_1\overleftarrow{\mathbf{x}}}(k, m) &= \left[\, 1 \;\rho_{X_1X_2}(k, m) \;\cdots\; \rho_{X_1X_N}(k, m)\,\right]^T\\ &= \frac{E\left[X_1(k, m)\overleftarrow{\mathbf{x}}^*(k, m)\right]}{E\left[|X_1(k, m)|^2\right]}\\ &= \overleftarrow{\mathbf{d}}^*(k)\end{aligned} \tag{4.9}$$

is the partially normalized [with respect to $X_1(k, m)$] correlation vector (of length N) between $X_1(k, m)$ and $\overleftarrow{\mathbf{x}}(k, m)$.

We see that $\overleftarrow{\mathbf{y}}(k, m)$ is the sum of two uncorrelated components. Therefore, the correlation matrix of $\overleftarrow{\mathbf{y}}(k, m)$ is

$$\begin{aligned}\boldsymbol{\Phi}_{\overleftarrow{\mathbf{y}}}(k, m) &= E\left[\overleftarrow{\mathbf{y}}(k, m)\overleftarrow{\mathbf{y}}^H(k, m)\right]\\ &= \phi_{X_1}(k, m)\boldsymbol{\rho}^*_{X_1\overleftarrow{\mathbf{x}}}(k, m)\boldsymbol{\rho}^T_{X_1\overleftarrow{\mathbf{x}}}(k, m) + \boldsymbol{\Phi}_{\overleftarrow{\mathbf{v}}}(k, m)\\ &= \boldsymbol{\Phi}_{\overleftarrow{\mathbf{x}}}(k, m) + \boldsymbol{\Phi}_{\overleftarrow{\mathbf{v}}}(k, m),\end{aligned} \tag{4.10}$$

where $\phi_{X_1}(k, m) = E\left[|X_1(k, m)|^2\right]$ is the variance of $X_1(k, m)$,

$$\boldsymbol{\Phi}_{\overleftarrow{\mathbf{x}}}(k, m) = \phi_{X_1}(k, m)\boldsymbol{\rho}^*_{X_1\overleftarrow{\mathbf{x}}}(k, m)\boldsymbol{\rho}^T_{X_1\overleftarrow{\mathbf{x}}}(k, m) \tag{4.11}$$

is the correlation matrix (whose rank is equal to 1) of $\overleftarrow{\mathbf{x}}(k, m)$, and

$$\boldsymbol{\Phi}_{\overleftarrow{\mathbf{v}}}(k, m) = E\left[\overleftarrow{\mathbf{v}}(k, m)\overleftarrow{\mathbf{v}}^H(k, m)\right] \tag{4.12}$$

is the correlation matrices of $\overleftarrow{\mathbf{v}}(k, m)$.

4.2 Array Signal Processing with Gains

In the STFT domain, the conventional beamforming or multichannel noise reduction is performed by applying a complex weight to the output of each sensor, at frequency-bin k, and summing across the aperture [3, 4]:

$$Z(k, m) = \overleftarrow{\mathbf{h}}^H(k, m)\overleftarrow{\mathbf{y}}(k, m), \tag{4.13}$$

where $\overleftarrow{\mathbf{h}}(k, m)$ is an FIR filter of length N containing all the complex gains applied to the microphone outputs at frequency-bin k.

We can express (4.13) as a function of the steering vector, i.e.,

$$\begin{aligned} Z(k, m) &= \overleftarrow{\mathbf{h}}^H(k, m)\left[\boldsymbol{\rho}^*_{X_1\overleftarrow{\mathbf{x}}}(k, m)X_1(k, m) + \overleftarrow{\mathbf{v}}(k, m)\right] \\ &= X_{\text{fd}}(k, m) + V_{\text{rn}}(k, m), \end{aligned} \tag{4.14}$$

where

$$X_{\text{fd}}(k, m) = X_1(k, m)\overleftarrow{\mathbf{h}}^H(k, m)\boldsymbol{\rho}^*_{X_1\overleftarrow{\mathbf{x}}}(k, m) \tag{4.15}$$

is the filtered desired signal and

$$V_{\text{rn}}(k, m) = \overleftarrow{\mathbf{h}}^H(k, m)\overleftarrow{\mathbf{v}}(k, m) \tag{4.16}$$

is the residual noise.

The two terms on the right-hand side of (4.14) are uncorrelated. Hence, the variance of $Z(k, m)$ is also the sum of two variances:

$$\begin{aligned} \phi_Z(k, m) &= \overleftarrow{\mathbf{h}}^H(k, m)\boldsymbol{\Phi}_{\overleftarrow{\mathbf{y}}}(k, m)\overleftarrow{\mathbf{h}}(k, m) \\ &= \phi_{X_{\text{fd}}}(k, m) + \phi_{V_{\text{rn}}}(k, m), \end{aligned} \tag{4.17}$$

where

$$\phi_{X_{\text{fd}}}(k, m) = \phi_{X_1}(k, m)\left|\overleftarrow{\mathbf{h}}^H(k, m)\boldsymbol{\rho}^*_{X_1\overleftarrow{\mathbf{x}}}(k, m)\right|^2, \tag{4.18}$$

$$\phi_{V_{\text{rn}}}(k, m) = \overleftarrow{\mathbf{h}}^H(k, m)\boldsymbol{\Phi}_{\overleftarrow{\mathbf{v}}}(k, m)\overleftarrow{\mathbf{h}}(k, m). \tag{4.19}$$

The different variances in (4.17) are important in the definitions of the performance measures.

The signal $X_1(k, m)$ is completely coherent across all sensors [see (4.5)]; however, $V_1(k, m)$ is usually partially coherent with the noise components, $V_n(k, m)$, at the other microphones. Therefore, the vector $\overleftarrow{\mathbf{v}}(k, m)$ can be decomposed into two orthogonal components, i.e.,

$$\overleftarrow{\mathbf{v}}(k, m) = \boldsymbol{\rho}^*_{V_1\overleftarrow{\mathbf{v}}}(k, m)V_1(k, m) + \overleftarrow{\mathbf{v}}_{\text{u}}(k, m), \tag{4.20}$$

where

$$\begin{aligned} \boldsymbol{\rho}_{V_1\overleftarrow{\mathbf{v}}}(k, m) &= \left[1 \ \rho_{V_1V_2}(k, m) \ \cdots \ \rho_{V_1V_N}(k, m)\right]^T \\ &= \frac{E\left[V_1(k, m)\overleftarrow{\mathbf{v}}^*(k, m)\right]}{E\left[|V_1(k, m)|^2\right]} \end{aligned} \tag{4.21}$$

is the partially normalized [with respect to $V_1(k, m)$] correlation vector (of length N) between $V_1(k, m)$ and $\overleftarrow{\mathbf{v}}(k, m)$,

$$\rho_{V_1 V_n}(k, m) = \frac{E\left[V_1(k, m)V_n^*(k, m)\right]}{E\left[|V_1(k, m)|^2\right]}, \quad n = 1, 2, \ldots, N \tag{4.22}$$

is the partially normalized [with respect to $V_1(k, m)$] correlation coefficient between $V_1(k, m)$ and $V_n(k, m)$, and $E\left[V_1^*(k, m)\overleftarrow{\mathbf{v}}_{\mathrm{u}}(k, m)\right] = \mathbf{0}$. Expression (4.14) is now

$$\begin{aligned} Z(k, m) &= \overleftarrow{\mathbf{h}}^H(k, m)\boldsymbol{\rho}^*_{X_1 \overleftarrow{\mathbf{x}}}(k, m)X_1(k, m) + \overleftarrow{\mathbf{h}}^H(k, m)\boldsymbol{\rho}^*_{V_1 \overleftarrow{\mathbf{v}}}(k, m)V_1(k, m) \\ &\quad + \overleftarrow{\mathbf{h}}^H(k, m)\overleftarrow{\mathbf{v}}_{\mathrm{u}}(k, m), \end{aligned} \tag{4.23}$$

which is the sum of three uncorrelated components. Thanks to this decomposition, more constraints are possible on $\overleftarrow{\mathbf{h}}(k, m)$.

4.3 Performance Measures

Since microphone 1 is our reference, all measures will be defined with respect to the signal from this microphone.

4.3.1 Noise Reduction

The subband and fullband input SNRs at time-frame m are defined as

$$\mathrm{iSNR}(k, m) = \frac{\phi_{X_1}(k, m)}{\phi_{V_1}(k, m)}, \quad k = 0, 1, \ldots, K-1, \tag{4.24}$$

$$\mathrm{iSNR}(m) = \frac{\sum_{k=0}^{K-1}\phi_{X_1}(k, m)}{\sum_{k=0}^{K-1}\phi_{V_1}(k, m)}, \tag{4.25}$$

where $\phi_{V_1}(k, m) = E\left[|V_1(k, m)|^2\right]$ is the variance of $V_1(k, m)$. It is easy to show that

$$\mathrm{iSNR}(m) \le \sum_{k=0}^{K-1} \mathrm{iSNR}(k, m). \tag{4.26}$$

The subband output SNR is obtained from (4.17):

$$\text{oSNR}\left[\overleftarrow{\mathbf{h}}(k,m)\right] = \frac{\phi_{X_{\text{fd}}}(k,m)}{\phi_{V_{\text{rn}}}(k,m)} = \frac{\phi_{X_1}(k,m)\left|\overleftarrow{\mathbf{h}}^H(k,m)\boldsymbol{\rho}^*_{X_1\overleftarrow{\mathbf{x}}}(k,m)\right|^2}{\overleftarrow{\mathbf{h}}^H(k,m)\boldsymbol{\Phi}_{\overleftarrow{\mathbf{v}}}(k,m)\overleftarrow{\mathbf{h}}(k,m)}, \; k = 0, 1, \ldots, K-1. \tag{4.27}$$

For the particular filter $\overleftarrow{\mathbf{h}}(k,m) = \mathbf{i}_{N,1}$, where $\mathbf{i}_{N,1}$ is the first column of the identity matrix $\mathbf{I}_N$ (of size $N \times N$), we have

$$\text{oSNR}\left[\mathbf{i}_{N,1}(k,m)\right] = \text{iSNR}(k,m). \tag{4.28}$$

With the identity filter, $\mathbf{i}_{N,1}$, the SNR cannot be improved.

For any two vectors $\overleftarrow{\mathbf{h}}(k,m)$ and $\boldsymbol{\rho}^*_{X_1\overleftarrow{\mathbf{x}}}(k,m)$ and a positive definite matrix $\boldsymbol{\Phi}_{\overleftarrow{\mathbf{v}}}(k,m)$, we have

$$\left|\overleftarrow{\mathbf{h}}^H(k,m)\boldsymbol{\rho}^*_{X_1\overleftarrow{\mathbf{x}}}(k,m)\right|^2 \le \left[\overleftarrow{\mathbf{h}}^H(k,m)\boldsymbol{\Phi}_{\overleftarrow{\mathbf{v}}}(k,m)\overleftarrow{\mathbf{h}}(k,m)\right] \times \left[\boldsymbol{\rho}^T_{X_1\overleftarrow{\mathbf{x}}}(k,m)\boldsymbol{\Phi}^{-1}_{\overleftarrow{\mathbf{v}}}(k,m)\boldsymbol{\rho}^*_{X_1\overleftarrow{\mathbf{x}}}(k,m)\right], \tag{4.29}$$

with equality if and only if $\overleftarrow{\mathbf{h}}(k,m) \propto \boldsymbol{\Phi}^{-1}_{\overleftarrow{\mathbf{v}}}(k,m)\boldsymbol{\rho}^*_{X_1\overleftarrow{\mathbf{x}}}(k,m)$. Using the previous inequality in (4.27), we deduce an upper bound for the subband output SNR:

$$\text{oSNR}\left[\overleftarrow{\mathbf{h}}(k,m)\right] \le \phi_{X_1}(k,m) \times \boldsymbol{\rho}^T_{X_1\overleftarrow{\mathbf{x}}}(k,m)\boldsymbol{\Phi}^{-1}_{\overleftarrow{\mathbf{v}}}(k,m)\boldsymbol{\rho}^*_{X_1\overleftarrow{\mathbf{x}}}(k,m), \; \forall \overleftarrow{\mathbf{h}}(k,m) \tag{4.30}$$

and, clearly,

$$\text{oSNR}\left[\mathbf{i}_{N,1}(k,m)\right] \le \phi_{X_1}(k,m) \cdot \boldsymbol{\rho}^T_{X_1\overleftarrow{\mathbf{x}}}(k,m)\boldsymbol{\Phi}^{-1}_{\overleftarrow{\mathbf{v}}}(k,m)\boldsymbol{\rho}^*_{X_1\overleftarrow{\mathbf{x}}}(k,m), \tag{4.31}$$

which implies that

$$\boldsymbol{\rho}^T_{X_1\overleftarrow{\mathbf{x}}}(k,m)\boldsymbol{\Phi}^{-1}_{\overleftarrow{\mathbf{v}}}(k,m)\boldsymbol{\rho}^*_{X_1\overleftarrow{\mathbf{x}}}(k,m) \ge \frac{1}{\phi_{V_1}(k,m)}. \tag{4.32}$$

The role of the beamformer is to produce a signal whose subband SNR is higher than that of the subband input SNR. This is measured by the subband array gain:

$$\mathcal{A}\left[\overleftarrow{\mathbf{h}}(k,m)\right] = \frac{\text{oSNR}\left[\overleftarrow{\mathbf{h}}(k,m)\right]}{\text{iSNR}(k,m)}, \; k = 0, 1, \ldots, K-1. \tag{4.33}$$

From (4.30), we deduce that the maximum subband array gain is

$$\mathcal{A}_{\max}(k,m) = \phi_{V_1}(k,m) \cdot \boldsymbol{\rho}^T_{X_1 \overleftarrow{\mathbf{x}}}(k,m) \boldsymbol{\Phi}^{-1}_{\overleftarrow{\mathbf{v}}}(k,m) \boldsymbol{\rho}^*_{X_1 \overleftarrow{\mathbf{x}}}(k,m) \geq 1. \tag{4.34}$$

We define the fullband output SNR at time-frame m as

$$\text{oSNR}\left[\overleftarrow{\mathbf{h}}(:,m)\right] = \frac{\sum_{k=0}^{K-1} \phi_{X_1}(k,m) \left| \overleftarrow{\mathbf{h}}^H(k,m) \boldsymbol{\rho}^*_{X_1 \overleftarrow{\mathbf{x}}}(k,m) \right|^2}{\sum_{k=0}^{K-1} \overleftarrow{\mathbf{h}}^H(k,m) \boldsymbol{\Phi}_{\overleftarrow{\mathbf{v}}}(k,m) \overleftarrow{\mathbf{h}}(k,m)} \tag{4.35}$$

and it can be verified that

$$\text{oSNR}\left[\overleftarrow{\mathbf{h}}(:,m)\right] \leq \sum_{k=0}^{K-1} \text{oSNR}\left[\overleftarrow{\mathbf{h}}(k,m)\right]. \tag{4.36}$$

Property 4.1 *We always have*

$$\text{oSNR}\left[\overleftarrow{\mathbf{h}}(:,m)\right] \leq \max_k \text{oSNR}\left[\overleftarrow{\mathbf{h}}(k,m)\right], \ \forall \overleftarrow{\mathbf{h}}(k,m). \tag{4.37}$$

Proof This is similar to what was shown in Chap. 3.

This property tells us that the fullband output SNR can never exceed the maximum subband output SNR for any filter $\overleftarrow{\mathbf{h}}(k,m)$.

We also define the fullband array gain as

$$\mathcal{A}\left[\overleftarrow{\mathbf{h}}(:,m)\right] = \frac{\text{oSNR}\left[\overleftarrow{\mathbf{h}}(:,m)\right]}{\text{iSNR}(m)}. \tag{4.38}$$

Other interesting measures for noise reduction are the subband and fullband noise reduction factors:

$$\begin{aligned}
\xi_{\text{nr}}\left[\overleftarrow{\mathbf{h}}(k,m)\right] &= \frac{\phi_{V_1}(k,m)}{\phi_{V_{\text{rn}}}(k,m)} \\
&= \frac{\phi_{V_1}(k,m)}{\overleftarrow{\mathbf{h}}^H(k,m) \boldsymbol{\Phi}_{\overleftarrow{\mathbf{v}}}(k,m) \overleftarrow{\mathbf{h}}(k,m)}, \ k = 0, 1, \ldots, K-1,
\end{aligned} \tag{4.39}$$

$$\xi_{\text{nr}}\left[\overleftarrow{\mathbf{h}}(:,m)\right] = \frac{\sum_{k=0}^{K-1} \phi_{V_1}(k,m)}{\sum_{k=0}^{K-1} \overleftarrow{\mathbf{h}}^H(k,m) \boldsymbol{\Phi}_{\overleftarrow{\mathbf{v}}}(k,m) \overleftarrow{\mathbf{h}}(k,m)}. \tag{4.40}$$

4.3.2 Speech Distortion

We can quantify the distortion of the desired signal via the subband and fullband speech reduction factors:

$$\xi_{\rm sr}\left[\overleftarrow{\mathbf{h}}(k,m)\right] = \frac{\phi_{X_1}(k,m)}{\phi_{X_{\rm fd}}(k,m)} = \frac{1}{\left|\overleftarrow{\mathbf{h}}^H(k,m)\boldsymbol{\rho}^*_{X_1\overleftarrow{\mathbf{x}}}(k,m)\right|^2},\ k = 0, 1, \ldots, K-1, \tag{4.41}$$

$$\xi_{\rm sr}\left[\overleftarrow{\mathbf{h}}(:,m)\right] = \frac{\sum_{k=0}^{K-1}\phi_{X_1}(k,m)}{\sum_{k=0}^{K-1}\phi_{X_1}(k,m)\left|\overleftarrow{\mathbf{h}}^H(k,m)\boldsymbol{\rho}^*_{X_1\overleftarrow{\mathbf{x}}}(k,m)\right|^2}, \tag{4.42}$$

or via the subband and fullband speech distortion indices:

$$\upsilon_{\rm sd}\left[\overleftarrow{\mathbf{h}}(k,m)\right] = \frac{E\left\{|X_{\rm fd}(k,m) - X_1(k,m)|^2\right\}}{\phi_{X_1}(k,m)} = \left|\overleftarrow{\mathbf{h}}^H(k,m)\boldsymbol{\rho}^*_{X_1\overleftarrow{\mathbf{x}}}(k,m) - 1\right|^2,\ k = 0, 1, \ldots, K-1, \tag{4.43}$$

$$\upsilon_{\rm sd}\left[\overleftarrow{\mathbf{h}}(:,m)\right] = \frac{\sum_{k=0}^{K-1}E\left\{|X_{\rm fd}(k,m) - X_1(k,m)|^2\right\}}{\sum_{k=0}^{K-1}\phi_{X_1}(k,m)}. \tag{4.44}$$

It is important to see that the design of a filter that does not distort the desired signal requires the constraint

$$\overleftarrow{\mathbf{h}}^H(k,m)\boldsymbol{\rho}^*_{X_1\overleftarrow{\mathbf{x}}}(k,m) = 1,\ \forall k, m. \tag{4.45}$$

We can verify the fundamental relations:

$$\mathcal{A}\left[\overleftarrow{\mathbf{h}}(k,m)\right] = \frac{\xi_{\rm nr}\left[\overleftarrow{\mathbf{h}}(k,m)\right]}{\xi_{\rm sr}\left[\overleftarrow{\mathbf{h}}(k,m)\right]},\ k = 0, 1, \ldots, K-1, \tag{4.46}$$

$$\mathcal{A}\left[\overleftarrow{\mathbf{h}}(:,m)\right] = \frac{\xi_{\rm nr}\left[\overleftarrow{\mathbf{h}}(:,m)\right]}{\xi_{\rm sr}\left[\overleftarrow{\mathbf{h}}(:,m)\right]}. \tag{4.47}$$

4.3.3 Other Measures

The beampattern is a convenient way to represent the response of the beamformer to the desired signal, $X_1(k,m)$, as a function of the steering vector $\boldsymbol{\rho}^*_{X_1\overleftarrow{\mathbf{x}}}(k,m)$ (or equivalently, the location of the source), assuming the absence of any noise or interference. This steering vector spans the ratios of acoustic impulse responses from any point in space to the array of sensors. Formally, the beampattern is defined as the ratio of the variance of the beamformer output when the source impinges with a steering vector $\boldsymbol{\rho}^*_{X_1 \overleftarrow{\mathbf{x}}}(k,m)$ to the variance of the desired signal [3]. From this definition, we deduce

- the fullband beampattern,

$$\mathcal{B}\left[\boldsymbol{\rho}^*_{X_1 \overleftarrow{\mathbf{x}}}(:,m)\right] = \frac{\sum_{k=0}^{K-1} \phi_{X_1}(k,m) \left| \overleftarrow{\mathbf{h}}^H(k,m)\boldsymbol{\rho}^*_{X_1 \overleftarrow{\mathbf{x}}}(k,m) \right|^2}{\sum_{k=0}^{K-1} \phi_{X_1}(k,m)}, \tag{4.48}$$

- and the subband beampattern,

$$\mathcal{B}\left[\boldsymbol{\rho}^*_{X_1 \overleftarrow{\mathbf{x}}}(k,m)\right] = \left| \overleftarrow{\mathbf{h}}^H(k,m)\boldsymbol{\rho}^*_{X_1 \overleftarrow{\mathbf{x}}}(k,m) \right|^2. \tag{4.49}$$

It is interesting to point out that the fullband beampattern is a linear combination of subband beampatterns:

$$\mathcal{B}\left[\boldsymbol{\rho}^*_{X_1 \overleftarrow{\mathbf{x}}}(:,m)\right] = \frac{\sum_{k=0}^{K-1} \phi_{X_1}(k,m) \mathcal{B}\left[\boldsymbol{\rho}^*_{X_1 \overleftarrow{\mathbf{x}}}(k,m)\right]}{\sum_{k=0}^{K-1} \phi_{X_1}(k,m)}, \tag{4.50}$$

as the denominator is simply a scaling factor. The contribution of each subband beampattern to the overall fullband beampattern is proportional to the power of the desired signal at that frequency.

It is also interesting to observe the following relations:

$$\mathcal{B}\left[\boldsymbol{\rho}^*_{X_1 \overleftarrow{\mathbf{x}}}(:,m)\right] = \frac{1}{\xi_{\mathrm{sr}}\left[\overleftarrow{\mathbf{h}}(:,m)\right]},$$

$$\mathcal{B}\left[\boldsymbol{\rho}^*_{X_1 \overleftarrow{\mathbf{x}}}(k,m)\right] = \frac{1}{\xi_{\mathrm{sr}}\left[\overleftarrow{\mathbf{h}}(k,m)\right]}.$$

When the weights of the beamformer are chosen in such a way that there is no cancellation, the value of the beampattern is 1 in the direction of the source.

Usually in room acoustics, there are different kind of noise sources present at the same time. In order to model this situation, a spherically isotropic or "diffuse" noise field is one in which the noise power is constant and equal at all spatial frequencies (i.e., directions) [4, 5]. When designing beamformers, one would like to be able to quantify the ability of the beamformer to attenuate such a noise field. To that end, the directivity factor is classically defined as the array gain of a (subband) beamformer in an isotropic noise field. In the subband case, this is equivalent to the ratio of the beampattern in the direction of the source over the resulting residual noise power. Thus, we define

- the fullband directivity factor,

$$\mathcal{D}(m) = \frac{\mathcal{B}\left[\boldsymbol{\rho}^*_{X_1 \overleftarrow{\mathbf{x}}}(:, m)\right]}{\sum_{k=0}^{K-1} \overleftarrow{\mathbf{h}}^H(k, m)\boldsymbol{\Gamma}_{\text{diff}}(k)\overleftarrow{\mathbf{h}}(k, m)}, \tag{4.51}$$

- and the subband directivity factor,

$$\mathcal{D}(k, m) = \frac{\mathcal{B}\left[\boldsymbol{\rho}^*_{X_1 \overleftarrow{\mathbf{x}}}(k, m)\right]}{\overleftarrow{\mathbf{h}}^H(k, m)\boldsymbol{\Gamma}_{\text{diff}}(k)\overleftarrow{\mathbf{h}}(k, m)}, \tag{4.52}$$

where

$$\begin{aligned}[\boldsymbol{\Gamma}_{\text{diff}}(k)]_{ij} &= \frac{\sin 2\pi k d_{ij} c^{-1} K^{-1}}{2\pi k d_{ij} c^{-1} K^{-1}} \\ &= \text{sinc}\left[2\pi k d_{ij} c^{-1} K^{-1}\right]\end{aligned} \tag{4.53}$$

is the coherence matrix of a diffuse noise field [6, 7], d_{ij} is the distance between sensors i and j, and c is the propagation speed of sound in the air.

The classical directivity index [4, 8] is simply

$$\mathcal{DI}(k, m) = 10 \log_{10} \mathcal{D}(k, m). \tag{4.54}$$

The subband white noise gain (WNG) is formally defined as the array gain with a spatially white noise field:

$$\begin{aligned}\mathcal{W}\left[\overleftarrow{\mathbf{h}}(k, m)\right] &= \frac{\left|\overleftarrow{\mathbf{h}}^H(k, m)\boldsymbol{\rho}^*_{X_1 \overleftarrow{\mathbf{x}}}(k, m)\right|^2}{\overleftarrow{\mathbf{h}}^H(k, m)\overleftarrow{\mathbf{h}}(k, m)} \\ &= \frac{\mathcal{B}\left[\boldsymbol{\rho}^*_{X_1 \overleftarrow{\mathbf{x}}}(k, m)\right]}{\overleftarrow{\mathbf{h}}^H(k, m)\overleftarrow{\mathbf{h}}(k, m)}.\end{aligned} \tag{4.55}$$

Analogously, we define the fullband WNG as

$$\mathcal{W}\left[\overleftarrow{\mathbf{h}}(:, m)\right] = \frac{\mathcal{B}\left[\boldsymbol{\rho}^*_{X_1 \overleftarrow{\mathbf{x}}}(:, m)\right]}{\sum_{k=0}^{K-1} \overleftarrow{\mathbf{h}}^H(k, m)\overleftarrow{\mathbf{h}}(k, m)}. \tag{4.56}$$

4.3.4 MSE Criterion

The error signal between the estimated and desired signals at the frequency-bin k and time-frame m is

$$\begin{aligned}\mathcal{E}(k,m) &= Z(k,m) - X_1(k,m)\\ &= \overleftarrow{\mathbf{h}}^H(k,m)\overleftarrow{\mathbf{y}}(k,m) - X_1(k,m),\end{aligned} \tag{4.57}$$

which can be rewritten as

$$\mathcal{E}(k,m) = \mathcal{E}_{\mathrm{d}}(k,m) + \mathcal{E}_{\mathrm{r}}(k,m), \tag{4.58}$$

where

$$\begin{aligned}\mathcal{E}_{\mathrm{d}}(k,m) &= X_{\mathrm{fd}}(k,m) - X_1(k,m)\\ &= \left[\overleftarrow{\mathbf{h}}^H(k,m)\boldsymbol{\rho}^*_{X_1\overleftarrow{\mathbf{x}}}(k,m) - 1\right]X_1(k,m)\end{aligned} \tag{4.59}$$

is the speech distortion due to the complex filter and

$$\begin{aligned}\mathcal{E}_{\mathrm{r}}(k,m) &= V_{\mathrm{rn}}(k,m)\\ &= \overleftarrow{\mathbf{h}}^H(k,m)\overleftarrow{\mathbf{v}}(k,m)\end{aligned} \tag{4.60}$$

represents the residual noise.

The subband MSE criterion is then

$$\begin{aligned}J\left[\overleftarrow{\mathbf{h}}(k,m)\right] &= E\left[|\mathcal{E}(k,m)|^2\right]\\ &= J_{\mathrm{d}}\left[\overleftarrow{\mathbf{h}}(k,m)\right] + J_{\mathrm{r}}\left[\overleftarrow{\mathbf{h}}(k,m)\right],\end{aligned} \tag{4.61}$$

where

$$\begin{aligned}J_{\mathrm{d}}\left[\overleftarrow{\mathbf{h}}(k,m)\right] &= E\left[|\mathcal{E}_{\mathrm{d}}(k,m)|^2\right]\\ &= E\left[|X_{\mathrm{fd}}(k,m) - X_1(k,m)|^2\right]\\ &= \phi_{X_1}(k,m)\left|\overleftarrow{\mathbf{h}}^H(k,m)\boldsymbol{\rho}^*_{X_1\overleftarrow{\mathbf{x}}}(k,m) - 1\right|^2\end{aligned} \tag{4.62}$$

and

$$\begin{aligned}J_{\mathrm{r}}\left[\overleftarrow{\mathbf{h}}(k,m)\right] &= E\left[|\mathcal{E}_{\mathrm{r}}(k,m)|^2\right]\\ &= E\left[|V_{\mathrm{rn}}(k,m)|^2\right]\\ &= \phi_{V_{\mathrm{rn}}}(k,m).\end{aligned} \tag{4.63}$$

For the two particular filters $\overleftarrow{\mathbf{h}}(k,m) = \mathbf{i}_{N,1}$ and $\overleftarrow{\mathbf{h}}(k,m) = \mathbf{0}_{N\times 1}$, we get

$$J\left[\mathbf{i}_{N,1}(k,m)\right] = J_{\mathrm{r}}\left[\mathbf{i}_{N,1}(k,m)\right] = \phi_{V_1}(k,m), \tag{4.64}$$

$$J\left[\mathbf{0}_{N\times 1}(k,m)\right] = J_{\mathrm{d}}\left[\mathbf{0}_{N\times 1}(k,m)\right] = \phi_{X_1}(k,m). \tag{4.65}$$

We then find that the subband NMSE with respect to $J\left[\mathbf{i}_{N,1}(k,m)\right]$ is

$$\begin{aligned}\widetilde{J}\left[\overleftarrow{\mathbf{h}}(k,m)\right] &= \frac{J\left[\overleftarrow{\mathbf{h}}(k,m)\right]}{J\left[\mathbf{i}_{N,1}(k,m)\right]} \\ &= \mathrm{iSNR}(k,m)\cdot \upsilon_{\mathrm{sd}}\left[\overleftarrow{\mathbf{h}}(k,m)\right] + \frac{1}{\xi_{\mathrm{nr}}\left[\overleftarrow{\mathbf{h}}(k,m)\right]},\end{aligned} \tag{4.66}$$

where

$$\upsilon_{\mathrm{sd}}\left[\overleftarrow{\mathbf{h}}(k,m)\right] = \frac{J_{\mathrm{d}}\left[\overleftarrow{\mathbf{h}}(k,m)\right]}{J_{\mathrm{d}}\left[\mathbf{0}_{N\times 1}(k,m)\right]}, \tag{4.67}$$

$$\xi_{\mathrm{nr}}\left[\overleftarrow{\mathbf{h}}(k,m)\right] = \frac{J_{\mathrm{r}}\left[\mathbf{i}_{N,1}(k,m)\right]}{J_{\mathrm{r}}\left[\overleftarrow{\mathbf{h}}(k,m)\right]}, \tag{4.68}$$

and the subband NMSE with respect to $J\left[\mathbf{0}_{N\times 1}(k,m)\right]$ is

$$\begin{aligned}\overline{J}\left[\overleftarrow{\mathbf{h}}(k,m)\right] &= \frac{J\left[\overleftarrow{\mathbf{h}}(k,m)\right]}{J\left[\mathbf{0}_{N\times 1}(k,m)\right]} \\ &= \upsilon_{\mathrm{sd}}\left[\overleftarrow{\mathbf{h}}(k,m)\right] + \frac{1}{\mathrm{oSNR}\left[\overleftarrow{\mathbf{h}}(k,m)\right]\cdot \xi_{\mathrm{sr}}\left[\overleftarrow{\mathbf{h}}(k,m)\right]}.\end{aligned} \tag{4.69}$$

We have

$$\widetilde{J}\left[\overleftarrow{\mathbf{h}}(k,m)\right] = \mathrm{iSNR}(k,m)\cdot \overline{J}\left[\overleftarrow{\mathbf{h}}(k,m)\right]. \tag{4.70}$$

We can also deduce that the fullband MSE and NMSEs at time-frame m are

$$\begin{aligned}J\left[\overleftarrow{\mathbf{h}}(:,m)\right] &= \frac{1}{K}\sum_{k=0}^{K-1} J\left[\overleftarrow{\mathbf{h}}(k,m)\right] \\ &= \frac{1}{K}\sum_{k=0}^{K-1} J_{\mathrm{d}}\left[\overleftarrow{\mathbf{h}}(k,m)\right] + \frac{1}{K}\sum_{k=0}^{K-1} J_{\mathrm{r}}\left[\overleftarrow{\mathbf{h}}(k,m)\right] \\ &= J_{\mathrm{d}}\left[\overleftarrow{\mathbf{h}}(:,m)\right] + J_{\mathrm{r}}\left[\overleftarrow{\mathbf{h}}(:,m)\right],\end{aligned} \tag{4.71}$$

$$\widetilde{J}\left[\overleftarrow{\mathbf{h}}(:,m)\right] = K\frac{J\left[\overleftarrow{\mathbf{h}}(:,m)\right]}{\sum_{k=0}^{K-1}\phi_{V_1}(k,m)}$$
$$= \text{iSNR}(m)\cdot\upsilon_{\text{sd}}\left[\overleftarrow{\mathbf{h}}(:,m)\right] + \frac{1}{\xi_{\text{nr}}\left[\overleftarrow{\mathbf{h}}(:,m)\right]}, \quad (4.72)$$

$$\overline{J}\left[\overleftarrow{\mathbf{h}}(:,m)\right] = K\frac{J\left[\overleftarrow{\mathbf{h}}(:,m)\right]}{\sum_{k=0}^{K-1}\phi_{X_1}(k,m)}$$
$$= \upsilon_{\text{sd}}\left[\overleftarrow{\mathbf{h}}(:,m)\right] + \frac{1}{\text{oSNR}\left[\overleftarrow{\mathbf{h}}(:,m)\right]\cdot\xi_{\text{sr}}\left[\overleftarrow{\mathbf{h}}(:,m)\right]}. \quad (4.73)$$

The objective of noise reduction in the STFT domain with a microphone array is to find the optimal gains, forming the filter $\overleftarrow{\mathbf{h}}(k,m)$, at each frequency-bin k and time-frame m that would either directly minimize $J\left[\overleftarrow{\mathbf{h}}(k,m)\right]$ or minimize $J_{\text{d}}\left[\overleftarrow{\mathbf{h}}(k,m)\right]$ or $J_{\text{r}}\left[\overleftarrow{\mathbf{h}}(k,m)\right]$ subject to some constraint.

4.4 Optimal Gains

We start this section by deriving the maximum SNR filter first, which is not derived from an MSE criterion.

4.4.1 Maximum SNR

Let us rewrite the subband output SNR:

$$\text{oSNR}\left[\overleftarrow{\mathbf{h}}(k,m)\right] = \frac{\phi_{X_1}(k,m)\overleftarrow{\mathbf{h}}^H(k,m)\boldsymbol{\rho}^*_{X_1\overleftarrow{\mathbf{x}}}(k,m)\boldsymbol{\rho}^T_{X_1\overleftarrow{\mathbf{x}}}(k,m)\overleftarrow{\mathbf{h}}(k,m)}{\overleftarrow{\mathbf{h}}^H(k,m)\boldsymbol{\Phi}_{\overleftarrow{\mathbf{v}}}(k,m)\overleftarrow{\mathbf{h}}(k,m)}. \quad (4.74)$$

The maximum SNR filter, $\overleftarrow{\mathbf{h}}_{\max}(k,m)$, is obtained by maximizing the subband output SNR as given above. In (4.74), we recognize the generalized Rayleigh quotient. It is well known that this quotient is maximized with the maximum eigenvector of the matrix $\phi_{X_1}(k,m)\boldsymbol{\Phi}^{-1}_{\overleftarrow{\mathbf{v}}}(k,m)\boldsymbol{\rho}^*_{X_1\overleftarrow{\mathbf{x}}}(k,m)\boldsymbol{\rho}^T_{X_1\overleftarrow{\mathbf{x}}}(k,m)$. Let us denote by $\lambda_{\max}(k,m)$ the maximum eigenvalue corresponding to this maximum eigenvector. Since the rank of the mentioned matrix is equal to 1, we have

$$\lambda_{\max}(k,m) = \text{tr}\left[\phi_{X_1}(k,m)\boldsymbol{\Phi}^{-1}_{\overleftarrow{\mathbf{v}}}(k,m)\boldsymbol{\rho}^*_{X_1\overleftarrow{\mathbf{x}}}(k,m)\boldsymbol{\rho}^T_{X_1\overleftarrow{\mathbf{x}}}(k,m)\right]$$
$$= \phi_{X_1}(k,m)\boldsymbol{\rho}^T_{X_1\overleftarrow{\mathbf{x}}}(k,m)\boldsymbol{\Phi}^{-1}_{\overleftarrow{\mathbf{v}}}(k,m)\boldsymbol{\rho}^*_{X_1\overleftarrow{\mathbf{x}}}(k,m). \quad (4.75)$$

As a result,

$$\text{oSNR}\left[\overleftarrow{\mathbf{h}}_{\max}(k,m)\right] = \lambda_{\max}(k,m), \tag{4.76}$$

which corresponds to the maximum possible subband output SNR and

$$\mathcal{A}\left[\overleftarrow{\mathbf{h}}_{\max}(k,m)\right] = \mathcal{A}_{\max}(k,m). \tag{4.77}$$

Let us denote by $\mathcal{A}_{\max}^{(n)}(k,m)$ the maximum subband array gain of a microphone array with n sensors. By virtue of the inclusion principle [9] for the matrix $\phi_{X_1}(k,m)\mathbf{\Phi}_{\overleftarrow{\mathbf{v}}}^{-1}(k,m)\boldsymbol{\rho}_{X_1\overleftarrow{\mathbf{x}}}^{*}(k,m)\boldsymbol{\rho}_{X_1\overleftarrow{\mathbf{x}}}^{T}(k,m)$, we have

$$\mathcal{A}_{\max}^{(N)}(k,m) \geq \mathcal{A}_{\max}^{(N-1)}(k,m) \geq \cdots \geq \mathcal{A}_{\max}^{(2)}(k,m) \geq \mathcal{A}_{\max}^{(1)}(k,m) = 1. \tag{4.78}$$

This shows that by increasing the number of microphones, we necessarily increase the gain. If there is one microphone only, the subband SNR cannot be improved as expected [1].

Obviously, we also have

$$\overleftarrow{\mathbf{h}}_{\max}(k,m) = \alpha(k,m)\mathbf{\Phi}_{\overleftarrow{\mathbf{v}}}^{-1}(k,m)\boldsymbol{\rho}_{X_1\overleftarrow{\mathbf{x}}}^{*}(k,m), \tag{4.79}$$

where $\alpha(k,m)$ is an arbitrary scaling factor different from zero. While this factor has no effect on the subband output SNR, it may have on the fullband output SNR and speech distortion. In fact, all filters (except for the LCMV) derived in the rest of this section are equivalent up to this scaling factor. These filters also try to find the respective scaling factors depending on what we optimize.

The fullband output SNR with the maximum SNR filter is

$$\text{oSNR}\left[\overleftarrow{\mathbf{h}}_{\max}(:,m)\right] = \frac{\sum_{k=0}^{K-1} \dfrac{|\alpha(k,m)|^2\,\lambda_{\max}^2(k,m)}{\phi_{X_1}(k,m)}}{\sum_{k=0}^{K-1} \dfrac{|\alpha(k,m)|^2\,\lambda_{\max}(k,m)}{\phi_{X_1}(k,m)}}. \tag{4.80}$$

We see that the performance (in terms of fullband SNR improvement) of the maximum SNR filter is quite dependent on the values of $\alpha(k,m)$.

4.4.2 Wiener

The Wiener filter is found by minimizing the subband MSE, $J\left[\overleftarrow{\mathbf{h}}(k,m)\right]$ [Eq. (4.61)]. We get

$$\begin{aligned}\overleftarrow{\mathbf{h}}_{\mathrm{W}}(k,m) &= \mathbf{\Phi}_{\overleftarrow{\mathbf{y}}}^{-1}(k,m)E\left[\overleftarrow{\mathbf{x}}(k,m)X_1^*(k,m)\right] \\ &= \phi_{X_1}(k,m)\mathbf{\Phi}_{\overleftarrow{\mathbf{y}}}^{-1}(k,m)\boldsymbol{\rho}_{X_1\overleftarrow{\mathbf{x}}}^{*}(k,m),\end{aligned} \tag{4.81}$$

that we can express as

$$\begin{aligned}\overleftarrow{\mathbf{h}}_{\mathrm{W}}(k,m) &= \mathbf{\Phi}_{\overleftarrow{\mathbf{y}}}^{-1}(k,m)\mathbf{\Phi}_{\overleftarrow{\mathbf{x}}}(k,m)\mathbf{i}_{N,1} \\ &= \left[\mathbf{I}_N - \mathbf{\Phi}_{\overleftarrow{\mathbf{y}}}^{-1}(k,m)\mathbf{\Phi}_{\overleftarrow{\mathbf{v}}}(k,m)\right]\mathbf{i}_{N,1}. \end{aligned} \tag{4.82}$$

In this case, the Wiener filter relies on the second-order statistics of the observation and noise signals.

We are now going to write the general form of the Wiener beamformer in another way that will make it easier to compare to other beamformers. We know that

$$\mathbf{\Phi}_{\overleftarrow{\mathbf{y}}}(k,m) = \phi_{X_1}(k,m)\boldsymbol{\rho}^*_{X_1\overleftarrow{\mathbf{x}}}(k,m)\boldsymbol{\rho}^T_{X_1\overleftarrow{\mathbf{x}}}(k,m) + \mathbf{\Phi}_{\overleftarrow{\mathbf{v}}}(k,m). \tag{4.83}$$

Determining the inverse of $\mathbf{\Phi}_{\overleftarrow{\mathbf{y}}}(k,m)$ from the previous expression with the Woodbury's identity, we get

$$\begin{aligned}\mathbf{\Phi}_{\overleftarrow{\mathbf{y}}}^{-1}(k,m) &= \mathbf{\Phi}_{\overleftarrow{\mathbf{v}}}^{-1}(k,m) \\ &\quad - \frac{\mathbf{\Phi}_{\overleftarrow{\mathbf{v}}}^{-1}(k,m)\boldsymbol{\rho}^*_{X_1\overleftarrow{\mathbf{x}}}(k,m)\boldsymbol{\rho}^T_{X_1\overleftarrow{\mathbf{x}}}(k,m)\mathbf{\Phi}_{\overleftarrow{\mathbf{v}}}^{-1}(k,m)}{\phi_{X_1}^{-1}(k,m) + \boldsymbol{\rho}^T_{X_1\overleftarrow{\mathbf{x}}}(k,m)\mathbf{\Phi}_{\overleftarrow{\mathbf{v}}}^{-1}(k,m)\boldsymbol{\rho}^*_{X_1\overleftarrow{\mathbf{x}}}(k,m)}. \end{aligned} \tag{4.84}$$

Substituting (4.84) into (4.81) gives

$$\overleftarrow{\mathbf{h}}_{\mathrm{W}}(k,m) = \frac{\phi_{X_1}(k,m)\mathbf{\Phi}_{\overleftarrow{\mathbf{v}}}^{-1}(k,m)\boldsymbol{\rho}^*_{X_1\overleftarrow{\mathbf{x}}}(k,m)}{1 + \phi_{X_1}(k,m)\boldsymbol{\rho}^T_{X_1\overleftarrow{\mathbf{x}}}(k,m)\mathbf{\Phi}_{\overleftarrow{\mathbf{v}}}^{-1}(k,m)\boldsymbol{\rho}^*_{X_1\overleftarrow{\mathbf{x}}}(k,m)}, \tag{4.85}$$

that we can rewrite as

$$\begin{aligned}\overleftarrow{\mathbf{h}}_{\mathrm{W}}(k,m) &= \frac{\mathbf{\Phi}_{\overleftarrow{\mathbf{v}}}^{-1}(k,m)\left[\mathbf{\Phi}_{\overleftarrow{\mathbf{y}}}(k,m) - \mathbf{\Phi}_{\overleftarrow{\mathbf{v}}}(k,m)\right]}{1 + \mathrm{tr}\left\{\mathbf{\Phi}_{\overleftarrow{\mathbf{v}}}^{-1}(k,m)\left[\mathbf{\Phi}_{\overleftarrow{\mathbf{y}}}(k,m) - \mathbf{\Phi}_{\overleftarrow{\mathbf{v}}}(k,m)\right]\right\}}\mathbf{i}_{N,1} \\ &= \frac{\mathbf{\Phi}_{\overleftarrow{\mathbf{v}}}^{-1}(k,m)\mathbf{\Phi}_{\overleftarrow{\mathbf{y}}}(k,m) - \mathbf{I}_N}{1 - N + \mathrm{tr}\left[\mathbf{\Phi}_{\overleftarrow{\mathbf{v}}}^{-1}(k,m)\mathbf{\Phi}_{\overleftarrow{\mathbf{y}}}(k,m)\right]}\mathbf{i}_{N,1}. \end{aligned} \tag{4.86}$$

From (4.85), we deduce that the subband output SNR is

$$\begin{aligned}\mathrm{oSNR}\left[\overleftarrow{\mathbf{h}}_{\mathrm{W}}(k,m)\right] &= \lambda_{\max}(k,m) \\ &= \mathrm{tr}\left[\mathbf{\Phi}_{\overleftarrow{\mathbf{v}}}^{-1}(k,m)\mathbf{\Phi}_{\overleftarrow{\mathbf{y}}}(k,m)\right] - N \end{aligned} \tag{4.87}$$

and, obviously,

$$\mathrm{oSNR}\left[\overleftarrow{\mathbf{h}}_{\mathrm{W}}(k,m)\right] \geq \mathrm{iSNR}(k,m), \tag{4.88}$$

since the Wiener filter maximizes the subband output SNR.

The speech distortion indices are

$$\upsilon_{\mathrm{sd}}\left[\overleftarrow{\mathbf{h}}_{\mathrm{W}}(k,m)\right]=\frac{1}{\left[1+\lambda_{\max}(k,m)\right]^2}, \tag{4.89}$$

$$\upsilon_{\mathrm{sd}}\left[\overleftarrow{\mathbf{h}}_{\mathrm{W}}(:,m)\right]=\frac{\sum_{k=0}^{K-1}\phi_{X_1}(k,m)\left[1+\lambda_{\max}(k,m)\right]^{-2}}{\sum_{k=0}^{K-1}\phi_{X_1}(k,m)}. \tag{4.90}$$

The higher the value of $\lambda_{\max}(k,m)$ (and/or the number of microphones), the less the desired signal is distorted.

It is also easy to find the noise reduction factors:

$$\xi_{\mathrm{nr}}\left[\overleftarrow{\mathbf{h}}_{\mathrm{W}}(k,m)\right]=\frac{\left[1+\lambda_{\max}(k,m)\right]^2}{\mathrm{iSNR}(k,m)\cdot\lambda_{\max}(k,m)}, \tag{4.91}$$

$$\xi_{\mathrm{nr}}\left[\overleftarrow{\mathbf{h}}_{\mathrm{W}}(:,m)\right]=\frac{\sum_{k=0}^{K-1}\phi_{X_1}(k,m)\mathrm{iSNR}^{-1}(k,m)}{\sum_{k=0}^{K-1}\phi_{X_1}(k,m)\lambda_{\max}(k,m)\left[1+\lambda_{\max}(k,m)\right]^{-2}}, \tag{4.92}$$

and the speech reduction factors:

$$\xi_{\mathrm{sr}}\left[\overleftarrow{\mathbf{h}}_{\mathrm{W}}(k,m)\right]=\frac{\left[1+\lambda_{\max}(k,m)\right]^2}{\lambda_{\max}^2(k,m)}, \tag{4.93}$$

$$\xi_{\mathrm{sr}}\left[\overleftarrow{\mathbf{h}}_{\mathrm{W}}(:,m)\right]=\frac{\sum_{k=0}^{K-1}\phi_{X_1}(k,m)}{\sum_{k=0}^{K-1}\phi_{X_1}(k,m)\lambda_{\max}^2(k,m)\left[1+\lambda_{\max}(k,m)\right]^{-2}}. \tag{4.94}$$

The fullband output SNR of the Wiener filter is

$$\mathrm{oSNR}\left[\overleftarrow{\mathbf{h}}_{\mathrm{W}}(:,m)\right]=\frac{\sum_{k=0}^{K-1}\phi_{X_1}(k,m)\dfrac{\lambda_{\max}^2(k,m)}{\left[1+\lambda_{\max}(k,m)\right]^2}}{\sum_{k=0}^{K-1}\phi_{X_1}(k,m)\dfrac{\lambda_{\max}(k,m)}{\left[1+\lambda_{\max}(k,m)\right]^2}}. \tag{4.95}$$

Property 4.2 *With the STFT-domain Wiener beamformer given in* (4.81), *the fullband output SNR is always greater than or equal to the fullband input SNR, i.e.,* $\mathrm{oSNR}\left[\overleftarrow{\mathbf{h}}_{\mathrm{W}}(:,m)\right]\geq\mathrm{iSNR}(m)$.

It is interesting to see that the two filters $\overleftarrow{\mathbf{h}}_{\mathrm{W}}(k,m)$ and $\overleftarrow{\mathbf{h}}_{\max}(k,m)$ differ only by a scaling factor. Indeed, taking

$$\alpha(k,m)=\frac{\phi_{X_1}(k,m)}{1+\lambda_{\max}(k,m)} \tag{4.96}$$

in (4.79) (maximum SNR filter), we find (4.85) (Wiener filter).

4.4.3 MVDR

The well-known MVDR beamformer proposed by Capon [10, 11] is easily derived by minimizing the subband MSE of the residual noise, $J_r\left[\overleftarrow{\mathbf{h}}(k,m)\right]$, with the constraint that the desired signal is not distorted. Mathematically, this is equivalent to

$$\min_{\overleftarrow{\mathbf{h}}(k,m)} \overleftarrow{\mathbf{h}}^H(k,m)\mathbf{\Phi}_{\overleftarrow{\mathbf{v}}}(k,m)\overleftarrow{\mathbf{h}}(k,m) \text{ subject to } \overleftarrow{\mathbf{h}}^H(k,m)\boldsymbol{\rho}^*_{X_1\overleftarrow{\mathbf{x}}}(k,m) = 1, \tag{4.97}$$

for which the solution is

$$\overleftarrow{\mathbf{h}}_{\text{MVDR}}(k,m) = \frac{\mathbf{\Phi}^{-1}_{\overleftarrow{\mathbf{v}}}(k,m)\boldsymbol{\rho}^*_{X_1\overleftarrow{\mathbf{x}}}(k,m)}{\boldsymbol{\rho}^T_{X_1\overleftarrow{\mathbf{x}}}(k,m)\mathbf{\Phi}^{-1}_{\overleftarrow{\mathbf{v}}}(k,m)\boldsymbol{\rho}^*_{X_1\overleftarrow{\mathbf{x}}}(k,m)}. \tag{4.98}$$

Using the fact that $\mathbf{\Phi}_{\overleftarrow{\mathbf{x}}}(k,m) = \phi_{X_1}(k,m)\boldsymbol{\rho}^*_{X_1\overleftarrow{\mathbf{x}}}(k,m)\boldsymbol{\rho}^T_{X_1\overleftarrow{\mathbf{x}}}(k,m)$, the explicit dependence of the above filter on the steering vector is eliminated to obtain the following forms:

$$\begin{aligned}\overleftarrow{\mathbf{h}}_{\text{MVDR}}(k,m) &= \frac{\mathbf{\Phi}^{-1}_{\overleftarrow{\mathbf{v}}}(k,m)\mathbf{\Phi}_{\overleftarrow{\mathbf{x}}}(k,m)}{\lambda_{\max}(k,m)}\mathbf{i}_{N,1} \\ &= \frac{\mathbf{\Phi}^{-1}_{\overleftarrow{\mathbf{v}}}(k,m)\mathbf{\Phi}_{\overleftarrow{\mathbf{y}}}(k,m) - \mathbf{I}_N}{\text{tr}\left[\mathbf{\Phi}^{-1}_{\overleftarrow{\mathbf{v}}}(k,m)\mathbf{\Phi}_{\overleftarrow{\mathbf{y}}}(k,m)\right] - N}\mathbf{i}_{N,1}.\end{aligned} \tag{4.99}$$

Obviously, we can also write the MVDR as

$$\overleftarrow{\mathbf{h}}_{\text{MVDR}}(k,m) = \frac{\mathbf{\Phi}^{-1}_{\overleftarrow{\mathbf{y}}}(k,m)\boldsymbol{\rho}^*_{X_1\overleftarrow{\mathbf{x}}}(k,m)}{\boldsymbol{\rho}^T_{X_1\overleftarrow{\mathbf{x}}}(k,m)\mathbf{\Phi}^{-1}_{\overleftarrow{\mathbf{y}}}(k,m)\boldsymbol{\rho}^*_{X_1\overleftarrow{\mathbf{x}}}(k,m)}. \tag{4.100}$$

Taking

$$\alpha(k,m) = \frac{\phi_{X_1}(k,m)}{\lambda_{\max}(k,m)} \tag{4.101}$$

in (4.79) (maximum SNR filter), we find (4.98) (MVDR filter), showing how the maximum SNR and MVDR filters are equivalent up to a scaling factor.

The Wiener and MVDR filters are simply related as follows

$$\overleftarrow{\mathbf{h}}_{\text{W}}(k,m) = C_{\text{W}}(k,m)\overleftarrow{\mathbf{h}}_{\text{MVDR}}(k,m), \tag{4.102}$$

where

$$\begin{aligned} C_{\mathrm{W}}(k,m) &= \overleftarrow{\mathbf{h}}_{\mathrm{W}}^{H}(k,m)\boldsymbol{\rho}^{*}_{X_1\overleftarrow{\mathbf{x}}}(k,m) \\ &= \frac{\lambda_{\max}(k,m)}{1+\lambda_{\max}(k,m)} \end{aligned} \tag{4.103}$$

can be seen as a single-channel STFT-domain Wiener gain. In fact, any filter of the form

$$\overleftarrow{\mathbf{h}}(k,m) = C(k,m)\overleftarrow{\mathbf{h}}_{\mathrm{MVDR}}(k,m), \tag{4.104}$$

where $C(k, m)$ is a real number, with $0 < C(k,m) < 1$, removes more noise than the MVDR filter at the price of some desired signal distortion, which is

$$\xi_{\mathrm{sr}}\left[\overleftarrow{\mathbf{h}}(k,m)\right] = \frac{1}{C^2(k,m)} \tag{4.105}$$

or

$$\upsilon_{\mathrm{sd}}\left[\overleftarrow{\mathbf{h}}(k,m)\right] = \left[C(k,m)-1\right]^2. \tag{4.106}$$

It can be verified that we always have

$$\mathrm{oSNR}\left[\overleftarrow{\mathbf{h}}_{\mathrm{MVDR}}(k,m)\right] = \mathrm{oSNR}\left[\overleftarrow{\mathbf{h}}_{\mathrm{W}}(k,m)\right], \tag{4.107}$$

$$\upsilon_{\mathrm{sd}}\left[\overleftarrow{\mathbf{h}}_{\mathrm{MVDR}}(k,m)\right] = 0, \tag{4.108}$$

$$\xi_{\mathrm{sr}}\left[\overleftarrow{\mathbf{h}}_{\mathrm{MVDR}}(k,m)\right] = 1, \tag{4.109}$$

and

$$\xi_{\mathrm{nr}}\left[\overleftarrow{\mathbf{h}}_{\mathrm{MVDR}}(k,m)\right] \le \xi_{\mathrm{nr}}\left[\overleftarrow{\mathbf{h}}_{\mathrm{W}}(k,m)\right], \tag{4.110}$$

$$\xi_{\mathrm{nr}}\left[\overleftarrow{\mathbf{h}}_{\mathrm{MVDR}}(:,m)\right] \le \xi_{\mathrm{nr}}\left[\overleftarrow{\mathbf{h}}_{\mathrm{W}}(:,m)\right]. \tag{4.111}$$

The MVDR beamformer rejects the maximum level of noise allowable without distorting the desired signal at each frequency.

While the subband output SNRs of the Wiener and MVDR are strictly equal, their fullband output SNRs are not. The fullband output SNR of the MVDR is

$$\mathrm{oSNR}\left[\overleftarrow{\mathbf{h}}_{\mathrm{MVDR}}(:,m)\right] = \frac{\sum_{k=0}^{K-1}\phi_{X_1}(k,m)}{\sum_{k=0}^{K-1}\phi_{X_1}(k,m)\lambda_{\max}^{-1}(k,m)} \tag{4.112}$$

and

$$\mathrm{oSNR}\left[\overleftarrow{\mathbf{h}}_{\mathrm{MVDR}}(:,m)\right] \le \mathrm{oSNR}\left[\overleftarrow{\mathbf{h}}_{\mathrm{W}}(:,m)\right]. \tag{4.113}$$

Property 4.3 *With the STFT-domain MVDR beamformer given in* (4.98)*, the fullband output SNR is always greater than or equal to the fullband input SNR, i.e.,* $\text{oSNR}\left[\overleftarrow{\mathbf{h}}_{\text{MVDR}}(:, m)\right] \geq \text{iSNR}(m)$.

4.4.4 Spatial Prediction

We know that

$$\overleftarrow{\mathbf{x}}(k, m) = \boldsymbol{\rho}^*_{X_1\overleftarrow{\mathbf{x}}}(k, m)X_1(k, m). \tag{4.114}$$

This means that $X_1(k, m)$ can be spatially predicted. Minimizing $\phi_Z(k, m)$ subject to the distortionless constraint, $\overleftarrow{\mathbf{h}}^H(k, m)\overleftarrow{\mathbf{x}}(k, m) = X_1(k, m)$, we find the spatial prediction filter

$$\overleftarrow{\mathbf{h}}_{\text{SP}}(k, m) = \frac{\boldsymbol{\Phi}^{-1}_{\overleftarrow{\mathbf{y}}}(k, m)\boldsymbol{\rho}^*_{X_1\overleftarrow{\mathbf{x}}}(k, m)}{\boldsymbol{\rho}^T_{X_1\overleftarrow{\mathbf{x}}}(k, m)\boldsymbol{\Phi}^{-1}_{\overleftarrow{\mathbf{y}}}(k, m)\boldsymbol{\rho}^*_{X_1\overleftarrow{\mathbf{x}}}(k, m)}, \tag{4.115}$$

which coincides with the MVDR filter.

4.4.5 Tradeoff

As we have learned from the previous sections, not much flexibility is associated with the Wiener and MVDR filters in the sense that we do not know in advance by how much the subband output SNR will be improved. However, in many practical situations, we wish to control the compromise between noise reduction and speech distortion, and the best way to do this is via the so-called tradeoff beamformer.

In the tradeoff approach, we minimize the subband speech distortion index with the constraint that the subband noise reduction factor is equal to a positive value that is greater than 1. Mathematically, this is equivalent to

$$\min_{\overleftarrow{\mathbf{h}}(k,m)} J_{\text{d}}\left[\overleftarrow{\mathbf{h}}(k, m)\right] \quad \text{subject to} \quad J_{\text{r}}\left[\overleftarrow{\mathbf{h}}(k, m)\right] = \beta J_{\text{r}}\left[\mathbf{i}_{N,1}(k, m)\right], \tag{4.116}$$

where $0 < \beta < 1$ to insure that we get some noise reduction. By using a Lagrange multiplier, $\mu > 0$, to adjoin the constraint to the cost function, we easily deduce the tradeoff filter

$$\begin{aligned}\overleftarrow{\mathbf{h}}_{\mathrm{T},\mu}(k,m) &= \phi_{X_1}(k,m)\left[\boldsymbol{\Phi}_{\overleftarrow{\mathbf{x}}}(k,m)+\mu\boldsymbol{\Phi}_{\overleftarrow{\mathbf{v}}}(k,m)\right]^{-1}\boldsymbol{\rho}^*_{X_1\overleftarrow{\mathbf{x}}}(k,m)\\ &= \frac{\phi_{X_1}(k,m)\boldsymbol{\Phi}^{-1}_{\overleftarrow{\mathbf{v}}}(k,m)\boldsymbol{\rho}^*_{X_1\overleftarrow{\mathbf{x}}}(k,m)}{\mu+\phi_{X_1}(k,m)\boldsymbol{\rho}^T_{X_1\overleftarrow{\mathbf{x}}}(k,m)\boldsymbol{\Phi}^{-1}_{\overleftarrow{\mathbf{v}}}(k,m)\boldsymbol{\rho}^*_{X_1\overleftarrow{\mathbf{x}}}(k,m)}\\ &= \frac{\boldsymbol{\Phi}^{-1}_{\overleftarrow{\mathbf{v}}}(k,m)\boldsymbol{\Phi}_{\overleftarrow{\mathbf{y}}}(k,m)-\mathbf{I}_N}{\mu-N+\mathrm{tr}\left[\boldsymbol{\Phi}^{-1}_{\overleftarrow{\mathbf{v}}}(k,m)\boldsymbol{\Phi}_{\overleftarrow{\mathbf{y}}}(k,m)\right]}\mathbf{i}_{N,1},\end{aligned} \tag{4.117}$$

where the Lagrange multiplier, μ, satisfies

$$J_{\mathrm{r}}\left[\overleftarrow{\mathbf{h}}_{\mathrm{T},\mu}(k,m)\right] = \beta J_{\mathrm{r}}\left[\mathbf{i}_{N,1}(k,m)\right]. \tag{4.118}$$

However, in practice it is not easy to determine the optimal μ. Therefore, when this parameter is chosen in an ad-hoc way, we can see that for

- $\mu = 1$, $\overleftarrow{\mathbf{h}}_{\mathrm{T},1}(k,m) = \overleftarrow{\mathbf{h}}_{\mathrm{W}}(k,m)$, which is the Wiener filter;
- $\mu = 0$, $\overleftarrow{\mathbf{h}}_{\mathrm{T},0}(k,m) = \overleftarrow{\mathbf{h}}_{\mathrm{MVDR}}(k,m)$, which is the MVDR filter;
- $\mu > 1$, results in a filter with low residual noise at the expense of high speech distortion;
- $\mu < 1$, results in a filter with high residual noise and low speech distortion.

Note that the MVDR cannot be derived from the first line of (4.117) since by taking $\mu = 0$, we have to invert a matrix that is not full rank.

It can be observed that the tradeoff, Wiener, and maximum SNR beamformers are equivalent up to a scaling factor. As a result, the subband output SNR of the tradeoff filter is independent of μ and is identical to the subband output SNR of the Wiener filter, i.e.,

$$\mathrm{oSNR}\left[\overleftarrow{\mathbf{h}}_{\mathrm{T},\mu}(k,m)\right] = \mathrm{oSNR}\left[\overleftarrow{\mathbf{h}}_{\mathrm{W}}(k,m)\right],\ \forall\mu \geq 0. \tag{4.119}$$

We have

$$\upsilon_{\mathrm{sd}}\left[\overleftarrow{\mathbf{h}}_{\mathrm{T},\mu}(k,m)\right] = \left[\frac{\mu}{\mu+\lambda_{\max}(k,m)}\right]^2, \tag{4.120}$$

$$\xi_{\mathrm{sr}}\left[\overleftarrow{\mathbf{h}}_{\mathrm{T},\mu}(k,m)\right] = \left[1+\frac{\mu}{\lambda_{\max}(k,m)}\right]^2, \tag{4.121}$$

$$\xi_{\mathrm{nr}}\left[\overleftarrow{\mathbf{h}}_{\mathrm{T},\mu}(k,m)\right] = \frac{\left[\mu+\lambda_{\max}(k,m)\right]^2}{\mathrm{iSNR}(k,m)\cdot\lambda_{\max}(k,m)}. \tag{4.122}$$

The tradeoff beamformer is interesting from several perspectives since it encompasses both the Wiener and MVDR filters. It is then useful to study the fullband output SNR and the fullband speech distortion index of the tradeoff filter.

It can be verified that the fullband output SNR of the tradeoff filter is

$$\text{oSNR}\left[\overleftarrow{\mathbf{h}}_{\text{T},\mu}(:,m)\right] = \frac{\sum_{k=0}^{K-1}\phi_{X_1}(k,m)\dfrac{\lambda_{\max}^2(k,m)}{[\mu+\lambda_{\max}(k,m)]^2}}{\sum_{k=0}^{K-1}\phi_{X_1}(k,m)\dfrac{\lambda_{\max}(k,m)}{[\mu+\lambda_{\max}(k,m)]^2}}. \tag{4.123}$$

Property 4.4 *The fullband output SNR of the STFT-domain tradeoff filter is an increasing function of the parameter* μ.

From Property 4.4, we deduce that the MVDR beamformer gives the smallest fullband output SNR.

While the fullband output SNR is upper bounded, it is easy to see that the fullband noise reduction factor and fullband speech reduction factor are not. So when μ goes to infinity, so are $\xi_{\text{nr}}\left[\overleftarrow{\mathbf{h}}_{\text{T},\mu}(:,m)\right]$ and $\xi_{\text{sr}}\left[\overleftarrow{\mathbf{h}}_{\text{T},\mu}(:,m)\right]$.

The fullband speech distortion index is

$$\upsilon_{\text{sd}}\left[\overleftarrow{\mathbf{h}}_{\text{T},\mu}(:,m)\right] = \frac{\sum_{k=0}^{K-1}\phi_{X_1}(k,m)\dfrac{\mu^2}{[\mu+\lambda_{\max}(k,m)]^2}}{\sum_{k=0}^{K-1}\phi_{X_1}(k,m)}. \tag{4.124}$$

Property 4.5 *The fullband speech distortion index of the STFT-domain tradeoff filter is an increasing function of the parameter* μ.

It is clear that

$$0 \le \upsilon_{\text{sd}}\left[\overleftarrow{\mathbf{h}}_{\text{T},\mu}(:,m)\right] \le 1, \ \forall\mu \ge 0. \tag{4.125}$$

Therefore, as μ increases, the fullband output SNR increases at the price of more distortion to the desired signal.

Property 4.6 *With the STFT-domain tradeoff beamformer given in* (4.117), *the fullband output SNR is always greater than or equal to the fullband input SNR, i.e.,* $\text{oSNR}\left[\overleftarrow{\mathbf{h}}_{\text{T},\mu}(:,m)\right] \ge \text{iSNR}(m), \ \forall\mu \ge 0$.

From the previous results, we deduce that for $\mu \ge 1$,

$$\text{oSNR}\left[\overleftarrow{\mathbf{h}}_{\text{MVDR}}(:,m)\right] \le \text{oSNR}\left[\overleftarrow{\mathbf{h}}_{\text{W}}(:,m)\right] \le \text{oSNR}\left[\overleftarrow{\mathbf{h}}_{\text{T},\mu}(:,m)\right], \tag{4.126}$$

$$\upsilon_{\text{sd}}\left[\overleftarrow{\mathbf{h}}_{\text{MVDR}}(:,m)\right] \le \upsilon_{\text{sd}}\left[\overleftarrow{\mathbf{h}}_{\text{W}}(:,m)\right] \le \upsilon_{\text{sd}}\left[\overleftarrow{\mathbf{h}}_{\text{T},\mu}(:,m)\right], \tag{4.127}$$

and for $0 \le \mu \le 1$,

$$\text{oSNR}\left[\overleftarrow{\mathbf{h}}_{\text{MVDR}}(:,m)\right] \le \text{oSNR}\left[\overleftarrow{\mathbf{h}}_{\text{T},\mu}(:,m)\right] \le \text{oSNR}\left[\overleftarrow{\mathbf{h}}_{\text{W}}(:,m)\right], \tag{4.128}$$

$$\upsilon_{\text{sd}}\left[\overleftarrow{\mathbf{h}}_{\text{MVDR}}(:,m)\right] \le \upsilon_{\text{sd}}\left[\overleftarrow{\mathbf{h}}_{\text{T},\mu}(:,m)\right] \le \upsilon_{\text{sd}}\left[\overleftarrow{\mathbf{h}}_{\text{W}}(:,m)\right]. \tag{4.129}$$

4.4.6 LCMV

To derive the LCMV beamformer, we are going to take advantage of the decomposition of the noise signal given in (4.20). The objective is to recover the desired signal, $X_1(k, m)$, (as in the MVDR) and completely remove the coherent components $V_1(k, m)\boldsymbol{\rho}^*_{V_1 \overleftarrow{\mathbf{v}}}(k, m)$. The two constraints can be put together in a matrix form as

$$\overleftarrow{\mathbf{C}}^H(k, m)\,\overleftarrow{\mathbf{h}}(k, m) = \begin{bmatrix} 1 \\ 0 \end{bmatrix}, \tag{4.130}$$

where

$$\overleftarrow{\mathbf{C}}(k, m) = \begin{bmatrix} \boldsymbol{\rho}^*_{X_1 \overleftarrow{\mathbf{x}}}(k, m) & \boldsymbol{\rho}^*_{V_1 \overleftarrow{\mathbf{v}}}(k, m) \end{bmatrix} \tag{4.131}$$

is our constraint matrix of size $N \times 2$.

We can express the correlation matrix of the noise as

$$\boldsymbol{\Phi}_{\overleftarrow{\mathbf{v}}}(k, m) = \boldsymbol{\Phi}_{\overleftarrow{\mathbf{v}}_{\mathrm{c}}}(k, m) + \boldsymbol{\Phi}_{\overleftarrow{\mathbf{v}}_{\mathrm{u}}}(k, m), \tag{4.132}$$

where

$$\boldsymbol{\Phi}_{\overleftarrow{\mathbf{v}}_{\mathrm{c}}}(k, m) = \phi_{V_1}(k, m)\boldsymbol{\rho}^*_{V_1 \overleftarrow{\mathbf{v}}}(k, m)\boldsymbol{\rho}^T_{V_1 \overleftarrow{\mathbf{v}}}(k, m) \tag{4.133}$$

and

$$\boldsymbol{\Phi}_{\overleftarrow{\mathbf{v}}_{\mathrm{u}}}(k, m) = E\left[\overleftarrow{\mathbf{v}}_{\mathrm{u}}(k, m)\,\overleftarrow{\mathbf{v}}^H_{\mathrm{u}}(k, m)\right]. \tag{4.134}$$

Then, our optimal filter is obtained by minimizing the energy of the residual uncorrelated noise, with the constraints that the coherent noise components are cancelled and the desired speech is preserved, i.e.,

$$\begin{aligned} \overleftarrow{\mathbf{h}}_{\mathrm{LCMV}}(k, m) &= \arg \min_{\overleftarrow{\mathbf{h}}_{\mathrm{LCMV}}(k,m)} \overleftarrow{\mathbf{h}}^H_{\mathrm{LCMV}}(k, m)\boldsymbol{\Phi}_{\overleftarrow{\mathbf{v}}_{\mathrm{u}}}(k, m)\,\overleftarrow{\mathbf{h}}_{\mathrm{LCMV}}(k, m) \\ &\quad \text{subject to} \quad \overleftarrow{\mathbf{C}}^H(k, m)\,\overleftarrow{\mathbf{h}}(k, m) = \begin{bmatrix} 1 \\ 0 \end{bmatrix}. \end{aligned} \tag{4.135}$$

The solution to (4.135) is given by

$$\overleftarrow{\mathbf{h}}_{\mathrm{LCMV}}(k, m) = \boldsymbol{\Phi}^{-1}_{\overleftarrow{\mathbf{v}}_{\mathrm{u}}}(k, m)\,\overleftarrow{\mathbf{C}}(k, m)\left[\overleftarrow{\mathbf{C}}^H(k, m)\boldsymbol{\Phi}^{-1}_{\overleftarrow{\mathbf{v}}_{\mathrm{u}}}(k, m)\,\overleftarrow{\mathbf{C}}(k, m)\right]^{-1}\begin{bmatrix} 1 \\ 0 \end{bmatrix}. \tag{4.136}$$

We always have

$$\text{oSNR}\left[\overleftarrow{\mathbf{h}}_{\text{LCMV}}(:, m)\right] \leq \text{oSNR}\left[\overleftarrow{\mathbf{h}}_{\text{MVDR}}(:, m)\right], \tag{4.137}$$

$$\upsilon_{\text{sd}}\left[\overleftarrow{\mathbf{h}}_{\text{LCMV}}(:, m)\right] = 0, \tag{4.138}$$

$$\xi_{\text{sr}}\left[\overleftarrow{\mathbf{h}}_{\text{LCMV}}(:, m)\right] = 1, \tag{4.139}$$

and

$$\xi_{\text{nr}}\left[\overleftarrow{\mathbf{h}}_{\text{LCMV}}(:, m)\right] \leq \xi_{\text{nr}}\left[\overleftarrow{\mathbf{h}}_{\text{MVDR}}(:, m)\right] \leq \xi_{\text{nr}}\left[\overleftarrow{\mathbf{h}}_{\text{W}}(:, m)\right]. \tag{4.140}$$

The LCMV filter can be an interesting solution in practical applications where the coherent noise is more problematic than the incoherent one.

We are now going to show an important relationship between the MVDR and LCMV beamformers [12].

It can be verified that the inverse of $\mathbf{\Phi}_{\overleftarrow{\mathbf{v}}}(k, m)$ is

$$\mathbf{\Phi}^{-1}_{\overleftarrow{\mathbf{v}}}(k, m) = \mathbf{\Phi}^{-1}_{\overleftarrow{\mathbf{v}}_{\text{u}}}(k, m) - \frac{\mathbf{\Phi}^{-1}_{\overleftarrow{\mathbf{v}}_{\text{u}}}(k, m)\mathbf{\Phi}_{\overleftarrow{\mathbf{v}}_{\text{c}}}(k, m)\mathbf{\Phi}^{-1}_{\overleftarrow{\mathbf{v}}_{\text{u}}}(k, m)}{1 + \text{tr}\left[\mathbf{\Phi}^{-1}_{\overleftarrow{\mathbf{v}}_{\text{u}}}(k, m)\mathbf{\Phi}_{\overleftarrow{\mathbf{v}}_{\text{c}}}(k, m)\right]}. \tag{4.141}$$

From the previous expression, we can rewrite the MVDR as

$$\begin{aligned}\overleftarrow{\mathbf{h}}_{\text{MVDR}}(k, m) &= \frac{\mathbf{\Phi}^{-1}_{\overleftarrow{\mathbf{v}}}(k, m)\mathbf{\Phi}_{\overleftarrow{\mathbf{x}}}(k, m)}{\text{tr}\left[\mathbf{\Phi}^{-1}_{\overleftarrow{\mathbf{v}}}(k, m)\mathbf{\Phi}_{\overleftarrow{\mathbf{x}}}(k, m)\right]}\mathbf{i}_{N,1} \\ &= \frac{\left\{1 + \text{tr}\left[\mathbf{\Phi}^{-1}_{\overleftarrow{\mathbf{v}}_{\text{u}}}(k, m)\mathbf{\Phi}_{\overleftarrow{\mathbf{v}}_{\text{c}}}(k, m)\right]\right\}\mathbf{I}_N - \mathbf{\Phi}^{-1}_{\overleftarrow{\mathbf{v}}_{\text{u}}}(k, m)\mathbf{\Phi}_{\overleftarrow{\mathbf{v}}_{\text{c}}}(k, m)}{\delta(k, m) + \text{tr}\left[\mathbf{\Phi}^{-1}_{\overleftarrow{\mathbf{v}}_{\text{u}}}(k, m)\mathbf{\Phi}_{\overleftarrow{\mathbf{x}}}(k, m)\right]} \\ &\quad \times \mathbf{\Phi}^{-1}_{\overleftarrow{\mathbf{v}}_{\text{u}}}(k, m)\mathbf{\Phi}_{\overleftarrow{\mathbf{x}}}(k, m)\mathbf{i}_{N,1},\end{aligned} \tag{4.142}$$

where

$$\begin{aligned}\delta(k, m) = &\ \text{tr}\left[\mathbf{\Phi}^{-1}_{\overleftarrow{\mathbf{v}}_{\text{u}}}(k, m)\mathbf{\Phi}_{\overleftarrow{\mathbf{v}}_{\text{c}}}(k, m)\right]\text{tr}\left[\mathbf{\Phi}^{-1}_{\overleftarrow{\mathbf{v}}_{\text{u}}}(k, m)\mathbf{\Phi}_{\overleftarrow{\mathbf{x}}}(k, m)\right] \\ &- \text{tr}\left[\mathbf{\Phi}^{-1}_{\overleftarrow{\mathbf{v}}_{\text{u}}}(k, m)\mathbf{\Phi}_{\overleftarrow{\mathbf{v}}_{\text{c}}}(k, m)\mathbf{\Phi}^{-1}_{\overleftarrow{\mathbf{v}}_{\text{u}}}(k, m)\mathbf{\Phi}_{\overleftarrow{\mathbf{x}}}(k, m)\right].\end{aligned} \tag{4.143}$$

When the coherent noise is zero then the MVDR reduces to the matched filter [12]

$$\overleftarrow{\mathbf{h}}_{\text{MATCH}}(k, m) = \frac{\mathbf{\Phi}^{-1}_{\overleftarrow{\mathbf{v}}_{\text{u}}}(k, m)\mathbf{\Phi}_{\overleftarrow{\mathbf{x}}}(k, m)}{\text{tr}\left[\mathbf{\Phi}^{-1}_{\overleftarrow{\mathbf{v}}_{\text{u}}}(k, m)\mathbf{\Phi}_{\overleftarrow{\mathbf{x}}}(k, m)\right]}\mathbf{i}_{N,1}. \tag{4.144}$$

On the other hand, it can be checked that the LCMV can be rewritten as

$$\overleftarrow{\mathbf{h}}_{\mathrm{LCMV}}(k,m) = \frac{\mathrm{tr}\left[\mathbf{\Phi}_{\overleftarrow{\mathbf{v}}_{\mathrm{u}}}^{-1}(k,m)\mathbf{\Phi}_{\overleftarrow{\mathbf{v}}_{\mathrm{c}}}(k,m)\right]\mathbf{I}_N - \mathbf{\Phi}_{\overleftarrow{\mathbf{v}}_{\mathrm{u}}}^{-1}(k,m)\mathbf{\Phi}_{\overleftarrow{\mathbf{v}}_{\mathrm{c}}}(k,m)}{\delta(k,m)} \times \mathbf{\Phi}_{\overleftarrow{\mathbf{v}}_{\mathrm{u}}}^{-1}(k,m)\mathbf{\Phi}_{\overleftarrow{\mathbf{x}}}(k,m)\mathbf{i}_{N,1}. \quad (4.145)$$

Using (4.142) and (4.145), we conclude that we have the following relationship [12]:

$$\overleftarrow{\mathbf{h}}_{\mathrm{MVDR}}(k,m) = \varrho_1(k,m)\overleftarrow{\mathbf{h}}_{\mathrm{LCMV}}(k,m) + \varrho_2(k,m)\overleftarrow{\mathbf{h}}_{\mathrm{MATCH}}(k,m), \quad (4.146)$$

where

$$\varrho_1(k,m) = \frac{\delta(k,m)}{\delta(k,m) + \mathrm{tr}\left[\mathbf{\Phi}_{\overleftarrow{\mathbf{v}}_{\mathrm{u}}}^{-1}(k,m)\mathbf{\Phi}_{\overleftarrow{\mathbf{x}}}(k,m)\right]}, \quad (4.147)$$

$$\varrho_2(k,m) = 1 - \varrho_1(k,m). \quad (4.148)$$

The relationship between the MVDR, LCMV, and matched filters has a very attractive form in which we see that the MVDR attempts to both reducing the uncorrelated noise by means of $\overleftarrow{\mathbf{h}}_{\mathrm{MATCH}}(k,m)$ and rejecting the coherent noise by means of $\overleftarrow{\mathbf{h}}_{\mathrm{LCMV}}(k,m)$ [12].

References

1. J. Benesty, J. Chen, Y. Huang, *Microphone Array Signal Processing* (Springer, Berlin, 2008)
2. M. Brandstein, D.B. Ward (eds.), *Microphone Arrays: Signal Processing Techniques and Applications* (Springer, Berlin, 2001)
3. J.P. Dmochowski, J. Benesty, Microphone arrays: fundamental concepts, in *Speech Processing in Modern Communication—Challenges and Perspectives*, ed. by I. Cohen, J. Benesty, S. Gannot (Springer, Berlin, 2010), Chapter 8: pp. 199–223
4. G.W. Elko, J. Meyer, Microphone arrays, in *Springer Handbook of Speech Processing*, ed. by J. Benesty, M.M. Sondhi, Y. Huang, (Springer, Berlin, 2008) Chapter 48: pp. 1021–1041
5. D.H. Johnson, D.E. Dudgeon, *Array Signal Processing—Concepts and Techniques* (Prentice-Hall, Englewood Cliffs, 1993)
6. I. A. McCowan H. Bourlard, Microphone array post-filter for diffuse noise field, in *Proceedings of IEEE ICASSP* (2002), pp. I-905–I-908
7. A. Spriet, Adaptive filtering techniques for noise reduction and acoustic feedback cancellation in hearing aids. Ph.D. thesis, Katholieke Universiteit Leuven, Belgium, 2004
8. W. Herbordt, Combination of robust adaptive beamforming with acoustic echo cancellation for acoustic human/machine interfaces. Ph.D. thesis, Erlangen–Nuremberg University, Germany, 2004
9. J.N. Franklin, *Matrix Theory*. (Prentice-Hall, Englewood Cliffs, 1968)

10. J. Capon, High resolution frequency-wavenumber spectrum analysis. Proc. IEEE **57**, 1408–1418 (1969)
11. R.T. Lacoss, Data adaptive spectral analysis methods. Geophysics **36**, 661–675 (1971)
12. M. Souden, J. Benesty, S. Affes, A study of the LCMV and MVDR noise reduction filters. IEEE Trans. Signal Process. **58**, 4925–4935 (2010)

Chapter 5
Multichannel Speech Enhancement with Filters

This chapter is a generalization of the previous one, where now the interframe correlation is taken into account in order to improve beamforming since speech signals are correlated at successive frames with the STFT. The signal model is the same as in Chap. 4, but the microphone array processing is different as explained below.

5.1 Array Signal Processing with Filters

Since the signals from successive frames are not necessarily uncorrelated, it is important to consider this information in the derivation of beamforming algorithms. As we did it in Chap. 2 but in the single-channel case, for multiple microphones the beamformer output is

$$\begin{aligned} Z(k,m) &= \sum_{l=0}^{L-1} \overleftarrow{\mathbf{h}}_l^H(k,m)\overleftarrow{\mathbf{y}}(k,m-l) \\ &= \underline{\mathbf{h}}^H(k,m)\underline{\mathbf{y}}(k,m), \end{aligned} \tag{5.1}$$

where L is the number of consecutive time-frames used for each one of the frequency-bins,[1] $\overleftarrow{\mathbf{h}}_l(k,m),\ l=0,1,\ldots,L-1$ are FIR filters of length N, and

$$\underline{\mathbf{h}}(k,m) = \left[\ \overleftarrow{\mathbf{h}}_0^T(k,m)\ \overleftarrow{\mathbf{h}}_1^T(k,m) \cdots \overleftarrow{\mathbf{h}}_{L-1}^T(k,m)\ \right]^T,$$

$$\underline{\mathbf{y}}(k,m) = \left[\ \overleftarrow{\mathbf{y}}^T(k,m)\ \overleftarrow{\mathbf{y}}^T(k,m-1) \cdots \overleftarrow{\mathbf{y}}^T(k,m-L+1)\ \right]^T,$$

are vectors of length NL. The case $L = 1$ corresponds, obviously, to the conventional STFT-domain linear beamforming [1] (see Chap. 4), the case $N = L = 1$

[1] We can use, if we like, different numbers of consecutive time-frames for different frequencies but to simplify the presentation, we stick with the same number L.

J. Benesty et al., *Speech Enhancement in the STFT Domain*,
SpringerBriefs in Electrical and Computer Engineering,
DOI: 10.1007/978-3-642-23250-3_5,

corresponds to the classical single-channel noise reduction in the STFT domain with a gain [2] (see Chap. 2), and the case $N = 1,\ L > 1$ is also the single-channel noise reduction in the STFT domain but with a filter where the interframe correlation is taken into account [3] (see Chap. 3).

Let us now decompose the signal $Z(k, m)$ into the following form:

$$\begin{aligned} Z(k,m) &= \underline{\mathbf{h}}^H(k,m)\underline{\mathbf{x}}(k,m) + \underline{\mathbf{h}}^H(k,m)\underline{\mathbf{v}}(k,m) \\ &= X_{1,\mathrm{f}}(k,m) + V_{\mathrm{rn}}(k,m), \end{aligned} \tag{5.2}$$

where

$$\underline{\mathbf{x}}(k,m) = \mathbf{x}_1(k,m) \otimes \overleftarrow{\mathbf{d}}(k), \tag{5.3}$$

$$\mathbf{x}_1(k,m) = \left[X_1(k,m) X_1(k,m-1) \cdots X_1(k,m-L+1)\right]^T,$$

$\otimes$ is the Kronecker product, $\overleftarrow{\mathbf{d}}(k)$ is defined in Chap. 4, $\underline{\mathbf{v}}(k,m)$ is defined in a similar way to $\underline{\mathbf{y}}(k,m)$,

$$X_{1,\mathrm{f}}(k,m) = \underline{\mathbf{h}}^H(k,m)\underline{\mathbf{x}}(k,m) \tag{5.4}$$

is a filtered version of the desired signal at L successive time-frames, and

$$V_{\mathrm{rn}}(k,m) = \underline{\mathbf{h}}^H(k,m)\underline{\mathbf{v}}(k,m) \tag{5.5}$$

is the residual noise which is uncorrelated with $X_{1,\mathrm{f}}(k,m)$.

At time-frame m, our desired signal is $X_1(k,m)$ [and not the whole vector $\underline{\mathbf{x}}(k,m)$ or $\mathbf{x}_1(k,m)$]. However, the vector $\underline{\mathbf{x}}(k,m)$ in $X_{1,\mathrm{f}}(k,m)$ contains both the desired signal, $X_1(k,m)$, and the components $X_1(k,m-l),\ l \neq 0$, which are not the desired signals at time-frame m but signals that are correlated with $X_1(k,m)$. Therefore, the elements $X_1(k,m-l),\ l \neq 0$, contain both a part of the desired signal and a component that we consider as an interference. This suggests that we should decompose $X_1(k,m-l)$ into two orthogonal components corresponding to the part of the desired signal and interference, i.e.,

$$X_1(k,m-l) = \rho^*_{X_1}(k,m,l)X_1(k,m) + X_{1,\mathrm{i}}(k,m-l), \tag{5.6}$$

where

$$X_{1,\mathrm{i}}(k,m-l) = X_1(k,m-l) - \rho^*_{X_1}(k,m,l)X_1(k,m), \tag{5.7}$$

$$E\left[X_1(k,m)X^*_{1,\mathrm{i}}(k,m-l)\right] = 0, \tag{5.8}$$

and

$$\rho_{X_1}(k,m,l) = \frac{E\left[X_1(k,m)X_1^*(k,m-l)\right]}{E\left[|X_1(k,m)|^2\right]} \tag{5.9}$$

is the interframe correlation coefficient of the signal $X_1(k,m)$. Hence, we can write the vector $\underline{\mathbf{x}}(k,m)$ as

$$\begin{aligned}\underline{\mathbf{x}}(k,m) &= X_1(k,m)\left[\boldsymbol{\rho}_{X_1}^*(k,m) \otimes \overleftarrow{\mathbf{d}}(k)\right] + \underline{\mathbf{x}}_{\mathrm{i}}(k,m) \\ &= X_1(k,m)\underline{\mathbf{d}}(k,m) + \underline{\mathbf{x}}_{\mathbf{i}}(k,m) \\ &= \underline{\mathbf{x}}_{\mathrm{d}}(k,m) + \underline{\mathbf{x}}_{\mathrm{i}}(k,m),\end{aligned} \tag{5.10}$$

where

$$\boldsymbol{\rho}_{X_1}(k,m) = \left[\, 1 \;\, \rho_{X_1}(k,m,1) \;\cdots\; \rho_{X_1}(k,m,L-1)\,\right]^T \tag{5.11}$$

is the (normalized) interframe correlation vector between $X_1(k,m)$ and $\mathbf{x}_1(k,m)$,

$$\underline{\mathbf{x}}_{\mathrm{i}}(k,m) = \left[X_{1,\mathrm{i}}(k,m)\, X_{1,\mathrm{i}}(k,m-1) \cdots X_{1,\mathrm{i}}(k,m-L+1)\right]^T \otimes \overleftarrow{\mathbf{d}}(k) \tag{5.12}$$

is the interference signal vector of length *NL*,

$$\underline{\mathbf{d}}(k,m) = \boldsymbol{\rho}_{X_1}^*(k,m) \otimes \overleftarrow{\mathbf{d}}(k) \tag{5.13}$$

is a vector of length *NL*, and

$$\underline{\mathbf{x}}_{\mathrm{d}}(k,m) = X_1(k,m)\underline{\mathbf{d}}(k,m) \tag{5.14}$$

is the desired signal vector. Using (5.10), we can rewrite (5.2) as

$$Z(k,m) = X_{\mathrm{fd}}(k,m) + X_{\mathrm{ri}}(k,m) + V_{\mathrm{rn}}(k,m), \tag{5.15}$$

where

$$X_{\mathrm{fd}}(k,m) = X_1(k,m)\underline{\mathbf{h}}^H(k,m)\underline{\mathbf{d}}(k,m) \tag{5.16}$$

is the filtered desired signal and

$$X_{\mathrm{ri}}(k,m) = \underline{\mathbf{h}}^H(k,m)\underline{\mathbf{x}}_{\mathrm{i}}(k,m) \tag{5.17}$$

is the residual interference. Note that the above decomposition of the signal $X_1(k,m-l)$ is critical in order to properly design optimal multichannel noise reduction filters with the interframe correlation scheme.

The three terms on the right-hand side of (5.15) are mutually uncorrelated. Therefore, the variance of $Z(k,m)$ is

$$\phi_Z(k,m) = \phi_{X_{\mathrm{fd}}}(k,m) + \phi_{X_{\mathrm{ri}}}(k,m) + \phi_{V_{\mathrm{rn}}}(k,m), \tag{5.18}$$

where

$$\begin{aligned}
\phi_{X_{\rm fd}}(k,m) &= E\left[|X_{\rm fd}(k,m)|^2\right] \\
&= \phi_{X_1}(k,m)\left|\underline{\mathbf{h}}^H(k,m)\underline{\mathbf{d}}(k,m)\right|^2, \\
&= \underline{\mathbf{h}}^H(k,m)\mathbf{\Phi}_{\underline{\mathbf{x}}_{\rm d}}(k,m)\underline{\mathbf{h}}(k,m) \qquad (5.19) \\
\phi_{X_{\rm ri}}(k,m) &= E\left[|X_{\rm ri}(k,m)|^2\right] \\
&= \underline{\mathbf{h}}^H(k,m)\mathbf{\Phi}_{\underline{\mathbf{x}}_{\rm i}}(k,m)\underline{\mathbf{h}}(k,m) \\
&= \underline{\mathbf{h}}^H(k,m)\mathbf{\Phi}_{\underline{\mathbf{x}}}(k,m)\underline{\mathbf{h}}(k,m) \\
&\quad - \phi_{X_1}(k,m)\left|\underline{\mathbf{h}}^H(k,m)\underline{\mathbf{d}}(k,m)\right|^2, \qquad (5.20) \\
\phi_{V_{\rm rn}}(k,m) &= E\left[|V_{\rm rn}(k,m)|^2\right] \\
&= \underline{\mathbf{h}}^H(k,m)\mathbf{\Phi}_{\underline{\mathbf{v}}}(k,m)\underline{\mathbf{h}}(k,m), \qquad (5.21)
\end{aligned}$$

$\mathbf{\Phi}_{\underline{\mathbf{x}}_{\rm d}}(k,m) = \phi_{X_1}(k,m)\underline{\mathbf{d}}(k,m)\underline{\mathbf{d}}^H(k,m)$ is the correlation matrix of $\underline{\mathbf{x}}_{\rm d}(k,m)$, with $\mathbf{\Phi}_{\underline{\mathbf{x}}}(k,m)$, $\mathbf{\Phi}_{\underline{\mathbf{x}}_{\rm i}}(k,m)$, and $\mathbf{\Phi}_{\underline{\mathbf{v}}}(k,m)$ being the correlation matrices of $\underline{\mathbf{x}}(k,m)$, $\underline{\mathbf{x}}_{\rm i}(k,m)$, and $\underline{\mathbf{v}}(k,m)$, respectively.

We can also decompose the noise vector, $\underline{\mathbf{v}}(k,m)$, into two orthogonal components as

$$\underline{\mathbf{v}}(k,m) = \boldsymbol{\rho}^*_{V_1\underline{\mathbf{v}}}(k,m)V_1(k,m) + \underline{\mathbf{v}}_{\rm u}(k,m), \qquad (5.22)$$

where

$$\boldsymbol{\rho}_{V_1\underline{\mathbf{v}}}(k,m) = \frac{E\left[V_1(k,m)\underline{\mathbf{v}}^*(k,m)\right]}{E\left[|V_1(k,m)|^2\right]} \qquad (5.23)$$

is the partially normalized [with respect to $V_1(k,m)$] correlation vector (of length NL) between $V_1(k,m)$ and $\underline{\mathbf{v}}(k,m)$ and $E\left[V_1^*(k,m)\underline{\mathbf{v}}_{\rm u}(k,m)\right] = \mathbf{0}$. Expression (5.2) is now

$$\begin{aligned}
Z(k,m) =& \underline{\mathbf{h}}^H(k,m)\underline{\mathbf{d}}(k,m)X_1(k,m) + \underline{\mathbf{h}}^H(k,m)\boldsymbol{\rho}^*_{V_1\underline{\mathbf{v}}}(k,m)V_1(k,m) \\
&+ \underline{\mathbf{h}}^H(k,m)\underline{\mathbf{v}}_{\mathbf{u}}(k,m), \qquad (5.24)
\end{aligned}$$

which is the sum of three uncorrelated components. Thanks to this decomposition, more constraints are possible on $\underline{\mathbf{h}}(k,m)$.

5.2 Performance Measures

Like in the previous chapter, all measures will be defined with respect to the reference microphone but in the context of interframe correlation.

5.2.1 Noise Reduction

The subband and fullband input SNRs were already defined in Chap. 4.

We define the subband output SNR as

$$\begin{aligned}\text{oSNR}\left[\underline{\mathbf{h}}(k,m)\right] &= \frac{\phi_{X_{\text{fd}}}(k,m)}{\phi_{X_{\text{ri}}}(k,m)+\phi_{V_{\text{rn}}}(k,m)} \\ &= \frac{\phi_{X_1}(k,m)\left|\underline{\mathbf{h}}^H(k,m)\underline{\mathbf{d}}(k,m)\right|^2}{\underline{\mathbf{h}}^H(k,m)\mathbf{\Phi}_{\text{in}}(k,m)\underline{\mathbf{h}}(k,m)},\end{aligned} \tag{5.25}$$

where

$$\mathbf{\Phi}_{\text{in}}(k,m) = \mathbf{\Phi}_{\underline{\mathbf{x}}_{\text{i}}}(k,m) + \mathbf{\Phi}_{\underline{\mathbf{v}}}(k,m) \tag{5.26}$$

is the interference-plus-noise correlation matrix. For the particular filter $\underline{\mathbf{h}}(k,m) = \mathbf{i}_{NL,1}$, where $\mathbf{i}_{NL,1}$ is the first column of the identity matrix $\mathbf{I}_{NL}$ (of size $NL \times NL$), we have

$$\text{oSNR}\left[\mathbf{i}_{NL,1}(k,m)\right] = \text{iSNR}(k,m). \tag{5.27}$$

With the identity filter, $\mathbf{i}_{NL,1}$, the SNR cannot be improved.

For any two vectors $\underline{\mathbf{h}}(k,m)$ and $\underline{\mathbf{d}}(k,m)$ and a positive definite matrix $\mathbf{\Phi}_{\text{in}}(k,m)$, we have

$$\begin{aligned}\left|\underline{\mathbf{h}}^H(k,m)\underline{\mathbf{d}}(k,m)\right|^2 \le &\left[\underline{\mathbf{h}}^H(k,m)\mathbf{\Phi}_{\text{in}}(k,m)\underline{\mathbf{h}}(k,m)\right] \\ &\times \left[\underline{\mathbf{d}}^H(k,m)\mathbf{\Phi}_{\text{in}}^{-1}(k,m)\underline{\mathbf{d}}(k,m)\right],\end{aligned} \tag{5.28}$$

with equality if and only if $\underline{\mathbf{h}}(k,m) \propto \mathbf{\Phi}_{\text{in}}^{-1}(k,m)\underline{\mathbf{d}}(k,m)$. Using the previous inequality in (5.25), we deduce an upper bound for the subband output SNR:

$$\text{oSNR}\left[\underline{\mathbf{h}}(k,m)\right] \le \phi_{X_1}(k,m)\cdot\underline{\mathbf{d}}^H(k,m)\mathbf{\Phi}_{\text{in}}^{-1}(k,m)\underline{\mathbf{d}}(k,m),\ \forall\underline{\mathbf{h}}(k,m) \tag{5.29}$$

and clearly

$$\text{oSNR}\left[\mathbf{i}_{NL,1}(k,m)\right] \le \phi_{X_1}(k,m)\cdot\underline{\mathbf{d}}^H(k,m)\mathbf{\Phi}_{\text{in}}^{-1}(k,m)\underline{\mathbf{d}}(k,m). \tag{5.30}$$

We define the subband array gain as

$$\mathcal{A}\left[\underline{\mathbf{h}}(k,m)\right] = \frac{\text{oSNR}\left[\underline{\mathbf{h}}(k,m)\right]}{\text{iSNR}(k,m)},\quad k=0,1,\ldots,K-1. \tag{5.31}$$

From (5.29), we deduce that maximum subband array gain is

$$\mathcal{A}_{\max}(k,m) = \phi_{V_1}(k,m)\cdot\underline{\mathbf{d}}^H(k,m)\mathbf{\Phi}_{\text{in}}^{-1}(k,m)\underline{\mathbf{d}}(k,m) \ge 1. \tag{5.32}$$

We define the fullband output SNR at time-frame m as

$$\text{oSNR}\left[\underline{\mathbf{h}}(:,m)\right] = \frac{\sum_{k=0}^{K-1} \phi_{X_1}(k,m) \left|\underline{\mathbf{h}}^H(k,m)\underline{\mathbf{d}}(k,m)\right|^2}{\sum_{k=0}^{K-1} \underline{\mathbf{h}}^H(k,m)\mathbf{\Phi}_{\text{in}}(k,m)\underline{\mathbf{h}}(k,m)} \tag{5.33}$$

and it can be shown that

$$\text{oSNR}\left[\underline{\mathbf{h}}(:,m)\right] \leq \max_k \ \text{oSNR}\left[\underline{\mathbf{h}}(k,m)\right], \ \forall \underline{\mathbf{h}}(k,m). \tag{5.34}$$

We also define the fullband array gain as

$$\mathcal{A}\left[\underline{\mathbf{h}}(:,m)\right] = \frac{\text{oSNR}\left[\underline{\mathbf{h}}(:,m)\right]}{\text{iSNR}(m)}. \tag{5.35}$$

We end this subsection by giving the subband and fullband noise reduction factors:

$$\begin{aligned}\xi_{\text{nr}}\left[\underline{\mathbf{h}}(k,m)\right] &= \frac{\phi_{V_1}(k,m)}{\phi_{V_{\text{rn}}}(k,m)} \\ &= \frac{\phi_{V_1}(k,m)}{\underline{\mathbf{h}}^H(k,m)\mathbf{\Phi}_{\text{in}}(k,m)\underline{\mathbf{h}}(k,m)}, \quad k = 0, 1, \ldots, K-1,\end{aligned} \tag{5.36}$$

$$\xi_{\text{nr}}\left[\underline{\mathbf{h}}(:,m)\right] = \frac{\sum_{k=0}^{K-1} \phi_{V_1}(k,m)}{\sum_{k=0}^{K-1} \underline{\mathbf{h}}^H(k,m)\mathbf{\Phi}_{\text{in}}(k,m)\underline{\mathbf{h}}(k,m)}. \tag{5.37}$$

5.2.2 Speech Distortion

We can quantify the distortion of the desired signal via the subband and fullband speech reduction factors:

$$\begin{aligned}\xi_{\text{sr}}\left[\underline{\mathbf{h}}(k,m)\right] &= \frac{\phi_{X_1}(k,m)}{\phi_{X_{\text{fd}}}(k,m)} \\ &= \frac{1}{\left|\underline{\mathbf{h}}^H(k,m)\underline{\mathbf{d}}(k,m)\right|^2}, \quad k = 0, 1, \ldots, K-1,\end{aligned} \tag{5.38}$$

$$\xi_{\text{sr}}\left[\underline{\mathbf{h}}(:,m)\right] = \frac{\sum_{k=0}^{K-1} \phi_{X_1}(k,m)}{\sum_{k=0}^{K-1} \phi_{X_1}(k,m) \left|\underline{\mathbf{h}}^H(k,m)\underline{\mathbf{d}}(k,m)\right|^2}, \tag{5.39}$$

or via the subband and fullband speech distortion indices:

$$\upsilon_{\text{sd}}\left[\underline{\mathbf{h}}(k,m)\right] = \frac{E\left\{|X_{\text{fd}}(k,m) - X_1(k,m)|^2\right\}}{\phi_{X_1}(k,m)}$$
$$= \left|\underline{\mathbf{h}}^H(k,m)\underline{\mathbf{d}}(k,m) - 1\right|^2, \quad k = 0, 1, \ldots, K-1, \tag{5.40}$$

$$\upsilon_{\text{sd}}\left[\underline{\mathbf{h}}(:,m)\right] = \frac{\sum_{k=0}^{K-1} E\left\{|X_{\text{fd}}(k,m) - X_1(k,m)|^2\right\}}{\sum_{k=0}^{K-1} \phi_{X_1}(k,m)}. \tag{5.41}$$

It is important to see that the design of a filter that does not distort the desired signal requires the constraint

$$\underline{\mathbf{h}}^H(k,m)\underline{\mathbf{d}}(k,m) = 1, \ \forall k, m. \tag{5.42}$$

We can verify the fundamental relations:

$$\mathcal{A}\left[\underline{\mathbf{h}}(k,m)\right] = \frac{\xi_{\text{nr}}\left[\underline{\mathbf{h}}(k,m)\right]}{\xi_{\text{sr}}\left[\underline{\mathbf{h}}(k,m)\right]}, \quad k = 0, 1, \ldots, K-1, \tag{5.43}$$
$$\mathcal{A}\left[\underline{\mathbf{h}}(:,m)\right] = \frac{\xi_{\text{nr}}\left[\underline{\mathbf{h}}(:,m)\right]}{\xi_{\text{sr}}\left[\underline{\mathbf{h}}(:,m)\right]}. \tag{5.44}$$

5.2.3 MSE Criterion

The error signal between the estimated and desired signals at the frequency-bin k and time-frame m is

$$\mathcal{E}(k,m) = Z(k,m) - X_1(k,m)$$
$$= \underline{\mathbf{h}}^H(k,m)\underline{\mathbf{y}}(k,m) - X_1(k,m), \tag{5.45}$$

which can be rewritten as the sum of two uncorrelated error signals:

$$\mathcal{E}(k,m) = \mathcal{E}_{\text{d}}(k,m) + \mathcal{E}_{\text{r}}(k,m), \tag{5.46}$$

where

$$\mathcal{E}_{\text{d}}(k,m) = X_{\text{fd}}(k,m) - X_1(k,m)$$
$$= \left[\underline{\mathbf{h}}^H(k,m)\underline{\mathbf{d}}(k,m) - 1\right] X_1(k,m) \tag{5.47}$$

is the speech distortion due to the complex filter and

$$\mathcal{E}_{\text{r}}(k,m) = X_{\text{ri}}(k,m) + V_{\text{rn}}(k,m)$$
$$= \underline{\mathbf{h}}^H(k,m)\underline{\mathbf{x}}_{\text{i}}(k,m) + \underline{\mathbf{h}}^H(k,m)\underline{\mathbf{v}}(k,m) \tag{5.48}$$

represents the residual interference-plus-noise.

The subband MSE criterion is then

$$J\left[\underline{\mathbf{h}}(k,m)\right] = E\left[|\mathcal{E}(k,m)|^2\right] = J_{\mathrm{d}}\left[\underline{\mathbf{h}}(k,m)\right] + J_{\mathrm{r}}\left[\underline{\mathbf{h}}(k,m)\right], \quad (5.49)$$

where

$$\begin{aligned} J_{\mathrm{d}}\left[\underline{\mathbf{h}}(k,m)\right] &= E\left[|\mathcal{E}_{\mathrm{d}}(k,m)|^2\right] \\ &= E\left[|X_{\mathrm{fd}}(k,m) - X_1(k,m)|^2\right] \\ &= \phi_{X_1}(k,m)\left|\underline{\mathbf{h}}^H(k,m)\underline{\mathbf{d}}(k,m) - 1\right|^2 \end{aligned} \quad (5.50)$$

and

$$\begin{aligned} J_{\mathrm{r}}\left[\underline{\mathbf{h}}(k,m)\right] &= E\left[|\mathcal{E}_{\mathrm{r}}(k,m)|^2\right] \\ &= E\left[|X_{\mathrm{ri}}(k,m)|^2\right] + E\left[|V_{\mathrm{rn}}(k,m)|^2\right] \\ &= \phi_{X_{\mathrm{ri}}}(k,m) + \phi_{V_{\mathrm{rn}}}(k,m). \end{aligned} \quad (5.51)$$

For the two particular filters $\underline{\mathbf{h}}(k,m) = \mathbf{i}_{NL,1}$ and $\underline{\mathbf{h}}(k,m) = \mathbf{0}_{NL\times 1}$, we get

$$J\left[\mathbf{i}_{NL,1}(k,m)\right] = J_{\mathrm{r}}\left[\mathbf{i}_{NL,1}(k,m)\right] = \phi_{V_1}(k,m), \quad (5.52)$$
$$J\left[\mathbf{0}_{NL\times 1}(k,m)\right] = J_{\mathrm{d}}\left[\mathbf{0}_{NL\times 1}(k,m)\right] = \phi_{X_1}(k,m). \quad (5.53)$$

We then find that the subband NMSE with respect to $J\left[\mathbf{i}_{NL,1}(k,m)\right]$ is

$$\begin{aligned} \widetilde{J}\left[\underline{\mathbf{h}}(k,m)\right] &= \frac{J\left[\underline{\mathbf{h}}(k,m)\right]}{J\left[\mathbf{i}_{NL,1}(k,m)\right]} \\ &= \mathrm{iSNR}(k,m)\cdot \upsilon_{\mathrm{sd}}\left[\underline{\mathbf{h}}(k,m)\right] + \frac{1}{\xi_{\mathrm{nr}}\left[\underline{\mathbf{h}}(k,m)\right]}, \end{aligned} \quad (5.54)$$

where

$$\upsilon_{\mathrm{sd}}\left[\underline{\mathbf{h}}(k,m)\right] = \frac{J_{\mathrm{d}}\left[\underline{\mathbf{h}}(k,m)\right]}{J_{\mathrm{d}}\left[\mathbf{0}_{NL\times 1}(k,m)\right]}, \quad (5.55)$$

$$\xi_{\mathrm{nr}}\left[\underline{\mathbf{h}}(k,m)\right] = \frac{J_{\mathrm{r}}\left[\mathbf{i}_{NL,1}(k,m)\right]}{J_{\mathrm{r}}\left[\underline{\mathbf{h}}(k,m)\right]}, \quad (5.56)$$

and the subband NMSE with respect to $J\left[\mathbf{0}_{NL\times 1}(k,m)\right]$ is

$$\begin{aligned} \overline{J}\left[\underline{\mathbf{h}}(k,m)\right] &= \frac{J\left[\underline{\mathbf{h}}(k,m)\right]}{J\left[\mathbf{0}_{NL\times 1}(k,m)\right]} \\ &= \upsilon_{\mathrm{sd}}\left[\underline{\mathbf{h}}(k,m)\right] + \frac{1}{\mathrm{oSNR}\left[\underline{\mathbf{h}}(k,m)\right]\cdot \xi_{\mathrm{sr}}\left[\underline{\mathbf{h}}(k,m)\right]}. \end{aligned} \quad (5.57)$$

We have

$$\widetilde{J}\left[\underline{\mathbf{h}}(k,m)\right] = \text{iSNR}(k,m) \cdot \overline{J}\left[\underline{\mathbf{h}}(k,m)\right]. \tag{5.58}$$

We can also deduce that the fullband MSE and NMSEs at time-frame m are

$$\begin{aligned} J\left[\underline{\mathbf{h}}(:,m)\right] &= \frac{1}{K}\sum_{k=0}^{K-1} J\left[\underline{\mathbf{h}}(k,m)\right] \\ &= \frac{1}{K}\sum_{k=0}^{K-1} J_{\text{d}}\left[\underline{\mathbf{h}}(k,m)\right] + \frac{1}{K}\sum_{k=0}^{K-1} J_{\text{r}}\left[\underline{\mathbf{h}}(k,m)\right] \\ &= J_{\text{d}}\left[\underline{\mathbf{h}}(:,m)\right] + J_{\text{r}}\left[\underline{\mathbf{h}}(:,m)\right], \end{aligned} \tag{5.59}$$

$$\begin{aligned} \widetilde{J}\left[\underline{\mathbf{h}}(:,m)\right] &= K\frac{J\left[\underline{\mathbf{h}}(:,m)\right]}{\sum_{k=0}^{K-1}\phi_{V_1}(k,m)} \\ &= \text{iSNR}(m)\cdot \upsilon_{\text{sd}}\left[\underline{\mathbf{h}}(:,m)\right] + \frac{1}{\xi_{\text{nr}}\left[\underline{\mathbf{h}}(:,m)\right]}, \end{aligned} \tag{5.60}$$

$$\begin{aligned} \overline{J}\left[\underline{\mathbf{h}}(:,m)\right] &= K\frac{J\left[\underline{\mathbf{h}}(:,m)\right]}{\sum_{k=0}^{K-1}\phi_{X_1}(k,m)} \\ &= \upsilon_{\text{sd}}\left[\underline{\mathbf{h}}(:,m)\right] + \frac{1}{\text{oSNR}\left[\underline{\mathbf{h}}(:,m)\right]\cdot \xi_{\text{sr}}\left[\underline{\mathbf{h}}(:,m)\right]}. \end{aligned} \tag{5.61}$$

5.3 Optimal Filters

In this section, we are briefly deriving the most important and useful filters.

5.3.1 Maximum SNR

By maximizing the subband output SNR, we find the maximum SNR filter:

$$\underline{\mathbf{h}}_{\max}(k,m) = \alpha(k,m)\mathbf{\Phi}_{\text{in}}^{-1}(k,m)\underline{\mathbf{d}}(k,m), \tag{5.62}$$

where $\alpha(k,m)$ is an arbitrary scaling factor different from zero. The subband output SNR corresponding to this filter is then

$$\text{oSNR}\left[\underline{\mathbf{h}}_{\max}(k,m)\right] = \lambda_{\max}(k,m), \tag{5.63}$$

where

$$\begin{aligned} \lambda_{\max}(k,m) &= \text{tr}\left[\phi_{X_1}(k,m)\mathbf{\Phi}_{\text{in}}^{-1}(k,m)\underline{\mathbf{d}}(k,m)\underline{\mathbf{d}}^H(k,m)\right] \\ &= \phi_{X_1}(k,m)\underline{\mathbf{d}}^H(k,m)\mathbf{\Phi}_{\text{in}}^{-1}(k,m)\underline{\mathbf{d}}(k,m). \end{aligned} \tag{5.64}$$

This value corresponds to the maximum possible subband output SNR and does not depend on $\alpha(k, m)$.

The fullband output SNR with the maximum SNR filter is

$$\text{oSNR}\left[\underline{\mathbf{h}}_{\max}(:, m)\right] = \frac{\sum_{k=0}^{K-1} \dfrac{|\alpha(k, m)|^2 \lambda_{\max}^2(k, m)}{\phi_{X_1}(k, m)}}{\sum_{k=0}^{K-1} \dfrac{|\alpha(k, m)|^2 \lambda_{\max}(k, m)}{\phi_{X_1}(k, m)}}. \tag{5.65}$$

We see that the performance (in terms of fullband SNR improvement) of the maximum SNR filter is quite dependent on the values of $\alpha(k, m)$.

5.3.2 Wiener

By minimizing the subband MSE, $J\left[\underline{\mathbf{h}}(k, m)\right]$ [Eq. (5.49)], we find the Wiener filter:

$$\begin{aligned} \underline{\mathbf{h}}_{\text{W}}(k, m) &= \phi_{X_1}(k, m)\mathbf{\Phi}_{\underline{\mathbf{y}}}^{-1}(k, m)\underline{\mathbf{d}}(k, m) \\ &= \mathbf{\Phi}_{\underline{\mathbf{y}}}^{-1}(k, m)\mathbf{\Phi}_{\underline{\mathbf{x}}}(k, m)\mathbf{i}_{NL,1} \\ &= \left[\mathbf{I}_{NL} - \mathbf{\Phi}_{\underline{\mathbf{y}}}^{-1}(k, m)\mathbf{\Phi}_{\underline{\mathbf{v}}}(k, m)\right]\mathbf{i}_{NL,1}. \end{aligned} \tag{5.66}$$

Another interesting way to express this filter is

$$\underline{\mathbf{h}}_{\text{W}}(k, m) = \frac{\phi_{X_1}(k, m)\mathbf{\Phi}_{\text{in}}^{-1}(k, m)\underline{\mathbf{d}}(k, m)}{1 + \phi_{X_1}(k, m)\underline{\mathbf{d}}^H(k, m)\mathbf{\Phi}_{\text{in}}^{-1}(k, m)\underline{\mathbf{d}}(k, m)}, \tag{5.67}$$

that we can rewrite as

$$\begin{aligned} \underline{\mathbf{h}}_{\text{W}}(k, m) &= \frac{\mathbf{\Phi}_{\text{in}}^{-1}(k, m)\left[\mathbf{\Phi}_{\underline{\mathbf{y}}}(k, m) - \mathbf{\Phi}_{\text{in}}(k, m)\right]}{1 + \text{tr}\left\{\mathbf{\Phi}_{\text{in}}^{-1}(k, m)\left[\mathbf{\Phi}_{\underline{\mathbf{y}}}(k, m) - \mathbf{\Phi}_{\text{in}}(k, m)\right]\right\}}\mathbf{i}_{NL,1} \\ &= \frac{\mathbf{\Phi}_{\text{in}}^{-1}(k, m)\mathbf{\Phi}_{\underline{\mathbf{y}}}(k, m) - \mathbf{I}_{NL}}{1 - NL + \text{tr}\left[\mathbf{\Phi}_{\text{in}}^{-1}(k, m)\mathbf{\Phi}_{\underline{\mathbf{y}}}(k, m)\right]}\mathbf{i}_{NL,1}. \end{aligned} \tag{5.68}$$

From (5.67), we deduce that the subband output SNR is

$$\begin{aligned} \text{oSNR}\left[\underline{\mathbf{h}}_{\text{W}}(k, m)\right] &= \lambda_{\max}(k, m) \\ &= \text{tr}\left[\mathbf{\Phi}_{\text{in}}^{-1}(k, m)\mathbf{\Phi}_{\underline{\mathbf{y}}}(k, m)\right] - NL \end{aligned} \tag{5.69}$$

and, obviously,

$$\text{oSNR}\left[\underline{\mathbf{h}}_{\text{W}}(k,m)\right] \geq \text{iSNR}(k,m), \tag{5.70}$$

since the Wiener filter maximizes the subband output SNR.

The speech distortion indices are

$$\upsilon_{\text{sd}}\left[\underline{\mathbf{h}}_{\text{W}}(k,m)\right] = \frac{1}{[1+\lambda_{\max}(k,m)]^2}, \tag{5.71}$$

$$\upsilon_{\text{sd}}\left[\underline{\mathbf{h}}_{\text{W}}(:,m)\right] = \frac{\sum_{k=0}^{K-1}\phi_{X_1}(k,m)\left[1+\lambda_{\max}(k,m)\right]^{-2}}{\sum_{k=0}^{K-1}\phi_{X_1}(k,m)}. \tag{5.72}$$

The higher the value of $\lambda_{\max}(k,m)$ (and/or the number of microphones), the less the desired signal is distorted.

It is also easy to find the noise reduction factors:

$$\xi_{\text{nr}}\left[\underline{\mathbf{h}}_{\text{W}}(k,m)\right] = \frac{[1+\lambda_{\max}(k,m)]^2}{\text{iSNR}(k,m)\cdot\lambda_{\max}(k,m)}, \tag{5.73}$$

$$\xi_{\text{nr}}\left[\underline{\mathbf{h}}_{\text{W}}(:,m)\right] = \frac{\sum_{k=0}^{K-1}\phi_{X_1}(k,m)\text{iSNR}^{-1}(k,m)}{\sum_{k=0}^{K-1}\phi_{X_1}(k,m)\lambda_{\max}(k,m)\left[1+\lambda_{\max}(k,m)\right]^{-2}}, \tag{5.74}$$

and the speech reduction factors:

$$\xi_{\text{sr}}\left[\underline{\mathbf{h}}_{\text{W}}(k,m)\right] = \frac{[1+\lambda_{\max}(k,m)]^2}{\lambda_{\max}^2(k,m)}, \tag{5.75}$$

$$\xi_{\text{sr}}\left[\underline{\mathbf{h}}_{\text{W}}(:,m)\right] = \frac{\sum_{k=0}^{K-1}\phi_{X_1}(k,m)}{\sum_{k=0}^{K-1}\phi_{X_1}(k,m)\lambda_{\max}^2(k,m)\left[1+\lambda_{\max}(k,m)\right]^{-2}}. \tag{5.76}$$

The fullband output SNR of the Wiener filter is

$$\text{oSNR}\left[\underline{\mathbf{h}}_{\text{W}}(:,m)\right] = \frac{\sum_{k=0}^{K-1}\phi_{X_1}(k,m)\dfrac{\lambda_{\max}^2(k,m)}{[1+\lambda_{\max}(k,m)]^2}}{\sum_{k=0}^{K-1}\phi_{X_1}(k,m)\dfrac{\lambda_{\max}(k,m)}{[1+\lambda_{\max}(k,m)]^2}}. \tag{5.77}$$

It can be shown that

$$\text{oSNR}\left[\underline{\mathbf{h}}_{\text{W}}(:,m)\right] \geq \text{iSNR}(m). \tag{5.78}$$

5.3.3 MVDR

By minimizing the subband MSE of the residual interference-plus-noise with the constraint that the desired signal is not distorted, we easily find the MVDR filter:

$$\underline{\mathbf{h}}_{\text{MVDR}}(k,m) = \frac{\phi_{X_1}(k,m)\boldsymbol{\Phi}_{\text{in}}^{-1}(k,m)\underline{\mathbf{d}}(k,m)}{\lambda_{\max}(k,m)} = \frac{\boldsymbol{\Phi}_{\text{in}}^{-1}(k,m)\boldsymbol{\Phi}_{\underline{\mathbf{y}}}(k,m) - \mathbf{I}_{NL}}{\text{tr}\left[\boldsymbol{\Phi}_{\text{in}}^{-1}(k,m)\boldsymbol{\Phi}_{\underline{\mathbf{y}}}(k,m)\right] - NL}\mathbf{i}_{NL,1}, \tag{5.79}$$

that we can rewrite as

$$\underline{\mathbf{h}}_{\text{MVDR}}(k,m) = \frac{\boldsymbol{\Phi}_{\underline{\mathbf{y}}}^{-1}(k,m)\underline{\mathbf{d}}(k,m)}{\underline{\mathbf{d}}^H(k,m)\boldsymbol{\Phi}_{\underline{\mathbf{y}}}^{-1}(k,m)\underline{\mathbf{d}}(k,m)}. \tag{5.80}$$

It is clear that we always have

$$\text{oSNR}\left[\underline{\mathbf{h}}_{\text{MVDR}}(k,m)\right] = \text{oSNR}\left[\underline{\mathbf{h}}_{\text{W}}(k,m)\right], \tag{5.81}$$

$$\upsilon_{\text{sd}}\left[\underline{\mathbf{h}}_{\text{MVDR}}(k,m)\right] = 0, \tag{5.82}$$

$$\xi_{\text{sr}}\left[\underline{\mathbf{h}}_{\text{MVDR}}(k,m)\right] = 1, \tag{5.83}$$

and

$$\xi_{\text{nr}}\left[\underline{\mathbf{h}}_{\text{MVDR}}(k,m)\right] \le \xi_{\text{nr}}\left[\underline{\mathbf{h}}_{\text{W}}(k,m)\right], \tag{5.84}$$

$$\xi_{\text{nr}}\left[\underline{\mathbf{h}}_{\text{MVDR}}(:,m)\right] \le \xi_{\text{nr}}\left[\underline{\mathbf{h}}_{\text{W}}(:,m)\right]. \tag{5.85}$$

The MVDR beamformer rejects the maximum level of noise allowable without distorting the desired signal at each frequency.

While the subband output SNRs of the Wiener and MVDR are strictly equal, their fullband output SNRs are not. The fullband output SNR of the MVDR is

$$\text{oSNR}\left[\underline{\mathbf{h}}_{\text{MVDR}}(:,m)\right] = \frac{\sum_{k=0}^{K-1}\phi_{X_1}(k,m)}{\sum_{k=0}^{K-1}\phi_{X_1}(k,m)\lambda_{\max}^{-1}(k,m)} \tag{5.86}$$

and

$$\text{oSNR}\left[\underline{\mathbf{h}}_{\text{MVDR}}(:,m)\right] \le \text{oSNR}\left[\underline{\mathbf{h}}_{\text{W}}(:,m)\right]. \tag{5.87}$$

It can be shown that

$$\text{oSNR}\left[\underline{\mathbf{h}}_{\text{MVDR}}(:,m)\right] \ge \text{iSNR}(m). \tag{5.88}$$

5.3.4 Spatio-Temporal Prediction

Assume that we can find a spatio-temporal prediction filter $\underline{\mathbf{h}}'(k,m)$ of length *NL* in such a way that

$$\underline{\mathbf{x}}(k,m) \approx X_1(k,m)\underline{\mathbf{h}}'(k,m). \tag{5.89}$$

The estimate of $X_1(k,m)$ becomes

$$Z(k,m) \approx X_1(k,m)\underline{\mathbf{h}}^H(k,m)\underline{\mathbf{h}}'(k,m) + \underline{\mathbf{h}}^H(k,m)\underline{\mathbf{v}}(k,m). \tag{5.90}$$

We can then derive a distortionless filter as follows:

$$\min_{\underline{\mathbf{h}}(k,m)} \phi_Z(k,m) \quad \text{subject to} \quad \underline{\mathbf{h}}^H(k,m)\underline{\mathbf{h}}'(k,m) = 1. \tag{5.91}$$

We easily deduce the solution

$$\underline{\mathbf{h}}_{\mathrm{P}}(k,m) = \frac{\mathbf{\Phi}_{\underline{\mathbf{y}}}^{-1}(k,m)\underline{\mathbf{h}}'(k,m)}{\underline{\mathbf{h}}'^H(k,m)\mathbf{\Phi}_{\underline{\mathbf{y}}}^{-1}(k,m)\underline{\mathbf{h}}'(k,m)}. \tag{5.92}$$

To find the optimal $\underline{\mathbf{h}}'(k,m)$ in the Wiener sense, we need to define the error signal vector

$$\underline{\mathbf{e}}_{\mathrm{P}}(k,m) = \underline{\mathbf{x}}(k,m) - X_1(k,m)\underline{\mathbf{h}}'(k,m) \tag{5.93}$$

and form the MSE

$$J\left[\underline{\mathbf{h}}'(k,m)\right] = E\left[\underline{\mathbf{e}}_{\mathrm{P}}^H(k,m)\underline{\mathbf{e}}_{\mathrm{P}}(k,m)\right]. \tag{5.94}$$

By minimizing $J\left[\underline{\mathbf{h}}'(k,m)\right]$ with respect to $\underline{\mathbf{h}}'(k,m)$, we obtain the filter

$$\underline{\mathbf{h}}'_{\mathrm{o}}(k,m) = \underline{\mathbf{d}}(k,m). \tag{5.95}$$

It is interesting to observe that the error signal vector with the optimal filter, $\underline{\mathbf{h}}'_{\mathrm{o}}(k,m)$, corresponds to the interference signal vector, i.e.,

$$\begin{aligned}\underline{\mathbf{e}}_{\mathrm{P,\,o}}(k,m) &= \underline{\mathbf{x}}(k,m) - X_1(k,m)\underline{\mathbf{d}}(k,m) \\ &= \underline{\mathbf{x}}_{\mathrm{i}}(k,m),\end{aligned} \tag{5.96}$$

which is expected from of the orthogonality principle.

Substituting (5.95) into (5.92), we find that

$$\underline{\mathbf{h}}_{\mathrm{P}}(k,m) = \frac{\mathbf{\Phi}_{\underline{\mathbf{y}}}^{-1}(k,m)\underline{\mathbf{d}}(k,m)}{\underline{\mathbf{d}}^H(k,m)\mathbf{\Phi}_{\underline{\mathbf{y}}}^{-1}(k,m)\underline{\mathbf{d}}(k,m)}. \tag{5.97}$$

Clearly, the two filters $\underline{\mathbf{h}}_{\mathrm{MVDR}}(k,m)$ and $\underline{\mathbf{h}}_{\mathrm{P}}(k,m)$ are identical.

It is possible to find another spatio-temporal prediction filter by optimizing the criteria

$$\min_{\underline{\mathbf{h}}(k,m)} \underline{\mathbf{h}}^H(k,m)\mathbf{\Phi}_{\underline{\mathbf{v}}}(k,m)\underline{\mathbf{h}}(k,m) \quad \text{subject to} \quad \underline{\mathbf{h}}^H(k,m)\underline{\mathbf{h}}'(k,m) = 1. \tag{5.98}$$

This leads to

$$\underline{\mathbf{h}}_{\mathrm{P},2}(k,m) = \frac{\mathbf{\Phi}_{\underline{\mathbf{v}}}^{-1}(k,m)\underline{\mathbf{d}}(k,m)}{\underline{\mathbf{d}}^H(k,m)\mathbf{\Phi}_{\underline{\mathbf{v}}}^{-1}(k,m)\underline{\mathbf{d}}(k,m)} \tag{5.99}$$

and, this time, this second spatio-temporal prediction filter is different from the MVDR filter.

5.3.5 Tradeoff

By minimizing the subband speech distortion index with the constraint that the subband noise reduction factor is equal to a positive value that is greater than 1, we find the tradeoff filter:

$$\underline{\mathbf{h}}_{\mathrm{T},\mu}(k,m) = \frac{\phi_{X_1}(k,m)\mathbf{\Phi}_{\mathrm{in}}^{-1}(k,m)\underline{\mathbf{d}}(k,m)}{\mu + \lambda_{\max}(k,m)}, \tag{5.100}$$

where $\mu \geq 0$. We observe that both the Wiener and MVDR filters are particular cases of the tradeoff filter.

It can be observed that the tradeoff, MVDR, Wiener, and maximum SNR beamformers are equivalent up to a scaling factor. As a result, the subband output SNR of the tradeoff filter is independent of μ and is identical to the subband output SNR of the Wiener filter, i.e.,

$$\mathrm{oSNR}\left[\underline{\mathbf{h}}_{\mathrm{T},\mu}(k,m)\right] = \mathrm{oSNR}\left[\underline{\mathbf{h}}_{\mathrm{W}}(k,m)\right], \ \forall \mu \geq 0. \tag{5.101}$$

We have

$$\upsilon_{\mathrm{sd}}\left[\underline{\mathbf{h}}_{\mathrm{T},\mu}(k,m)\right] = \left[\frac{\mu}{\mu + \lambda_{\max}(k,m)}\right]^2, \tag{5.102}$$

$$\xi_{\mathrm{sr}}\left[\underline{\mathbf{h}}_{\mathrm{T},\mu}(k,m)\right] = \left[1 + \frac{\mu}{\lambda_{\max}(k,m)}\right]^2, \tag{5.103}$$

$$\xi_{\mathrm{nr}}\left[\underline{\mathbf{h}}_{\mathrm{T},\mu}(k,m)\right] = \frac{[\mu + \lambda_{\max}(k,m)]^2}{\mathrm{iSNR}(k,m) \cdot \lambda_{\max}(k,m)}. \tag{5.104}$$

It can be verified that the fullband output SNR of the tradeoff filter is

$$\text{oSNR}\left[\underline{\mathbf{h}}_{\text{T},\mu}(:,m)\right] = \frac{\sum_{k=0}^{K-1} \phi_{X_1}(k,m)\dfrac{\lambda_{\max}^2(k,m)}{[\mu+\lambda_{\max}(k,m)]^2}}{\sum_{k=0}^{K-1} \phi_{X_1}(k,m)\dfrac{\lambda_{\max}(k,m)}{[\mu+\lambda_{\max}(k,m)]^2}} \quad (5.105)$$

and

$$\text{oSNR}\left[\underline{\mathbf{h}}_{\text{T},\mu}(:,m)\right] \geq \text{iSNR}(m), \ \forall \mu \geq 0. \quad (5.106)$$

5.3.6 LCMV

It is possible to derive the LCMV beamformer by exploiting the structure of the noise as given in (5.22). Since we want to perfectly recover the desired signal, $X_1(k,m)$, and completely remove the correlated components, $V_1(k,m)\boldsymbol{\rho}^*_{V_1\underline{\mathbf{v}}}(k,m)$, these two constraints can be put together in a matrix form as

$$\underline{\mathbf{C}}^H(k,m)\underline{\mathbf{h}}(k,m) = \begin{bmatrix} 1 \\ 0 \end{bmatrix}, \quad (5.107)$$

where

$$\underline{\mathbf{C}}(k,m) = \left[\underline{\mathbf{d}}(k,m)\boldsymbol{\rho}^*_{V_1\underline{\mathbf{v}}}(k,m)\right] \quad (5.108)$$

is our constraint matrix of size $NL \times 2$. Then, our optimal filter is obtained by minimizing the energy at the filter output, with the constraints that the correlated noise components are cancelled and the desired speech is preserved, i.e.,

$$\underline{\mathbf{h}}_{\text{LCMV}}(k,m) = \arg\min_{\underline{\mathbf{h}}(k,m)} \underline{\mathbf{h}}^H(k,m)\boldsymbol{\Phi}_{\underline{\mathbf{y}}}(k,m)\underline{\mathbf{h}}(k,m)$$
$$\text{subject to} \ \ \underline{\mathbf{C}}^H(k,m)\underline{\mathbf{h}}(k,m) = \begin{bmatrix} 1 \\ 0 \end{bmatrix}. \quad (5.109)$$

The solution to (5.109) is given by

$$\underline{\mathbf{h}}_{\text{LCMV}}(k,m) = \boldsymbol{\Phi}_{\underline{\mathbf{y}}}^{-1}(k,m)\underline{\mathbf{C}}(k,m)\left[\underline{\mathbf{C}}^H(k,m)\boldsymbol{\Phi}_{\underline{\mathbf{y}}}^{-1}(k,m)\underline{\mathbf{C}}(k,m)\right]^{-1}\begin{bmatrix} 1 \\ 0 \end{bmatrix}. \quad (5.110)$$

We have

$$\text{oSNR}\left[\underline{\mathbf{h}}_{\text{LCMV}}(k,m)\right] \leq \text{oSNR}\left[\underline{\mathbf{h}}_{\text{MVDR}}(k,m)\right], \tag{5.111}$$

$$\upsilon_{\text{sd}}\left[\underline{\mathbf{h}}_{\text{LCMV}}(k,m)\right] = 0, \tag{5.112}$$

$$\xi_{\text{sr}}\left[\underline{\mathbf{h}}_{\text{LCMV}}(k,m)\right] = 1, \tag{5.113}$$

and

$$\xi_{\text{nr}}\left[\underline{\mathbf{h}}_{\text{LCMV}}(k,m)\right] \leq \xi_{\text{nr}}\left[\underline{\mathbf{h}}_{\text{MVDR}}(k,m)\right] \leq \xi_{\text{nr}}\left[\underline{\mathbf{h}}_{\text{W}}(k,m)\right]. \tag{5.114}$$

The LCMV filter is able to remove all the correlated noise but at the price that its overall noise reduction is lower than that of the MVDR filter.

References

1. J. Benesty, J. Chen, Y. Huang, *Microphone Array Signal Processing* (Springer, Berlin, 2008)
2. J. Benesty, J. Chen, Y. Huang, I. Cohen, *Noise Reduction in Speech Processing* (Springer, Berlin, 2009)
3. J. Benesty, Y. Huang, *A Perspective on Single-Channel Frequency-Domain Speech Enhancement* (Morgan & Claypool Publishers, 2011)

Chapter 6
The Bifrequency Spectrum in Speech Enhancement

In this chapter, we explain how the bifrequency spectrum is introduced in the derivation of speech enhancement filters considering that the signals may be nonstationary. The focus is on the single-channel case where the interframe correlation is not taken into account. Generalization to other cases is straightforward.

6.1 Problem Formulation

Let $a(t)$ be a zero-mean random variable for which its frequency-domain representation is $A(k, m)$. We define the bifrequency spectrum as [1, 2]

$$\phi_A(k_1, k_2, m) = E\left[A(k_1, m)A^*(k_2, m)\right], \tag{6.1}$$

where k_1 and k_2 are possibly two different frequency-bins. Basically, the bifrequency spectrum is a measure of the correlation between two different frequencies of the same signal. If $a(t)$ is a wide-sense stationary signal then the bifrequency spectrum reduces to

$$\phi_A(k_1, k_2, m) = \begin{cases} \phi_A(k, m), & k_1 = k_2 = k \\ 0, & k_1 \neq k_2 \end{cases}. \tag{6.2}$$

Thus, for a stationary random process, two distinct Fourier coefficients are uncorrelated. However, for a nonstationary random process, the bifrequency spectrum will exhibit non-zero correlations along the so-called support curves other than the main diagonal $k_1 = k_2$.

It is well known that speech is nonstationary. Therefore, it seems appropriate when deriving speech enhancement algorithms in the STFT domain to take into account the spectral correlation that may not be negligible in this context (see Chap. 1). As a consequence, the well-known musical tone problem will likely disappear.

J. Benesty et al., *Speech Enhancement in the STFT Domain*,
SpringerBriefs in Electrical and Computer Engineering,
DOI: 10.1007/978-3-642-23250-3_6,

Let us consider the desired signal at all frequency-bins in a vector of length K:

$$\mathbf{x}(m) = \left[\, X(0,m)\; X(1,m)\; \cdots\; X(K-1,m) \,\right]^T. \tag{6.3}$$

We can estimate $\mathbf{x}(m)$ with

$$\mathbf{z}(m) = \mathbf{H}(m)\mathbf{y}(m), \tag{6.4}$$

where

$$\mathbf{H}(m) = \begin{bmatrix} \mathbf{h}_0^H(m) \\ \mathbf{h}_1^H(m) \\ \vdots \\ \mathbf{h}_{K-1}^H(m) \end{bmatrix} \tag{6.5}$$

is a square filtering matrix of size $K \times K$, $\mathbf{h}_k(m)$, $k = 0, 1, \ldots, K-1$ are FIR filters of length K, and $\mathbf{y}(m)$ is defined in a similar way to $\mathbf{x}(m)$. We can rewrite (6.4) as

$$\begin{aligned} \mathbf{z}(m) &= \mathbf{H}(m)\left[\mathbf{x}(m) + \mathbf{v}(m)\right] \\ &= \mathbf{x}_{\text{fd}}(m) + \mathbf{v}_{\text{rn}}(m), \end{aligned} \tag{6.6}$$

where $\mathbf{v}(m)$ is defined similarly to $\mathbf{x}(m)$,

$$\mathbf{x}_{\text{fd}}(m) = \mathbf{H}(m)\mathbf{x}(m) \tag{6.7}$$

is the filtered desired signal vector, and

$$\mathbf{v}_{\text{rn}}(m) = \mathbf{H}(m)\mathbf{v}(m) \tag{6.8}$$

is the residual noise signal vector.

The correlation matrix of $\mathbf{z}(m)$ is then

$$\begin{aligned} \mathbf{\Phi}_{\mathbf{z}}(m) &= E\left[\mathbf{z}(m)\mathbf{z}^H(m)\right] \\ &= \mathbf{H}(m)\mathbf{\Phi}_{\mathbf{y}}(m)\mathbf{H}^H(m) \\ &= \mathbf{H}(m)\mathbf{\Phi}_{\mathbf{x}}(m)\mathbf{H}^H(m) + \mathbf{H}(m)\mathbf{\Phi}_{\mathbf{v}}(m)\mathbf{H}^H(m), \end{aligned} \tag{6.9}$$

where

$$\mathbf{\Phi}_{\mathbf{y}}(m) = \begin{bmatrix} \phi_Y(0,m) & \phi_Y(0,1,m) & \cdots & \phi_Y(0,K-1,m) \\ \phi_Y(1,0,m) & \phi_Y(1,m) & \cdots & \phi_Y(1,K-1,m) \\ \vdots & \vdots & \ddots & \vdots \\ \phi_Y(K-1,0,m) & \phi_Y(K-1,1,m) & \cdots & \phi_Y(K-1,m) \end{bmatrix}, \tag{6.10}$$

and $\mathbf{\Phi}_{\mathbf{x}}(m)$ and $\mathbf{\Phi}_{\mathbf{v}}(m)$ are defined in a similar way to $\mathbf{\Phi}_{\mathbf{y}}(m)$. We see now that the spectral correlation is taken into account in the estimator $\mathbf{z}(m)$. If the spectral correlation is negligible for both the speech and noise, then all three correlation matrices $\mathbf{\Phi_y}(m)$, $\mathbf{\Phi_x}(m)$, and $\mathbf{\Phi_v}(m)$ are diagonal and this approach is identical to the microphone signal processing with a gain explained in Chap. 2. In practice, only a very few off diagonals on both sides of the main diagonal of $\mathbf{H}(m)$ should be considered while the others could be neglected. This reasonable approximation will significantly reduce the complexity of this method. Also, we observe that if the STFT can be approximated by the Fourier matrix, then (6.6) is equivalent to the time-domain approach where K successive samples of the desired signal are estimated at each block [3].

6.2 Performance Measures

From the formulation shown in Sect. 6.1, it makes sense to discuss fullband performance measures only.

6.2.1 Noise Reduction

The fullband input SNR at time-frame m is

$$\mathrm{iSNR}(m) = \frac{\mathrm{tr}\left[\mathbf{\Phi_x}(m)\right]}{\mathrm{tr}\left[\mathbf{\Phi_v}(m)\right]}, \tag{6.11}$$

which is, obviously, equivalent to the one defined in Chap. 2.

From (6.9), we deduce that the fullband output SNR at time-frame m is

$$\mathrm{oSNR}\left[\mathbf{H}(m)\right] = \frac{\mathrm{tr}\left[\mathbf{H}(m)\mathbf{\Phi_x}(m)\mathbf{H}^H(m)\right]}{\mathrm{tr}\left[\mathbf{H}(m)\mathbf{\Phi_v}(m)\mathbf{H}^H(m)\right]}. \tag{6.12}$$

For the particular filtering matrix $\mathbf{H}(m) = \mathbf{I}_K$, where $\mathbf{I}_K$ is the identity matrix of size $K \times K$, we have

$$\mathrm{oSNR}\left[\mathbf{I}_K(m)\right] = \mathrm{iSNR}(m). \tag{6.13}$$

In this scenario, the SNR cannot be improved.

The fullband noise reduction factor is defined as

$$\xi_{\mathrm{nr}}\left[\mathbf{H}(m)\right] = \frac{\mathrm{tr}\left[\mathbf{\Phi_v}(m)\right]}{\mathrm{tr}\left[\mathbf{H}(m)\mathbf{\Phi_v}(m)\mathbf{H}^H(m)\right]}. \tag{6.14}$$

This factor should be greater than 1 for optimal filtering matrices.

6.2.2 Speech Distortion

We define the fullband speech reduction factor as

$$\xi_{\mathrm{sr}}\left[\mathbf{H}(m)\right] = \frac{\mathrm{tr}\left[\mathbf{\Phi}_{\mathbf{x}}(m)\right]}{\mathrm{tr}\left[\mathbf{H}(m)\mathbf{\Phi}_{\mathbf{x}}(m)\mathbf{H}^{H}(m)\right]}, \tag{6.15}$$

which should be upper bounded by 1.

It is clear that

$$\frac{\mathrm{oSNR}\left[\mathbf{H}(m)\right]}{\mathrm{iSNR}(m)} = \frac{\xi_{\mathrm{nr}}\left[\mathbf{H}(m)\right]}{\xi_{\mathrm{sr}}\left[\mathbf{H}(m)\right]}. \tag{6.16}$$

Another interesting way to quantify the distortion of the desired speech signal is with the fullband speech distortion index:

$$\begin{aligned} \upsilon_{\mathrm{sd}}\left[\mathbf{H}(m)\right] &= \frac{E\left\{\left[\mathbf{x}_{\mathrm{fd}}(m) - \mathbf{x}(m)\right]^{H}\left[\mathbf{x}_{\mathrm{fd}}(m) - \mathbf{x}(m)\right]\right\}}{\mathrm{tr}\left[\mathbf{\Phi}_{\mathbf{x}}(m)\right]} \\ &= \frac{\mathrm{tr}\left\{\left[\mathbf{H}(m) - \mathbf{I}_K\right]\mathbf{\Phi}_{\mathbf{x}}(m)\left[\mathbf{H}(m) - \mathbf{I}_K\right]^{H}\right\}}{\mathrm{tr}\left[\mathbf{\Phi}_{\mathbf{x}}(m)\right]}. \end{aligned} \tag{6.17}$$

For optimal filtering matrices, we should have $\upsilon_{\mathrm{sd}}\left[\mathbf{H}(m)\right] \le 1$.

6.2.3 MSE Criterion

Since the desired signal is a vector of length K, so is the error signal. We define the error signal vector between the estimated and desired signals as

$$\begin{aligned} \mathbf{e}(m) &= \mathbf{z}(m) - \mathbf{x}(m) \\ &= \mathbf{H}(m)\mathbf{y}(m) - \mathbf{x}(m), \end{aligned} \tag{6.18}$$

which can also be written as the sum of two orthogonal error signal vectors:

$$\mathbf{e}(m) = \mathbf{e}_{\mathrm{d}}(m) + \mathbf{e}_{\mathrm{r}}(m), \tag{6.19}$$

where

$$\mathbf{e}_{\mathrm{d}}(m) = \left[\mathbf{H}(m) - \mathbf{I}_K\right]\mathbf{x}(m) \tag{6.20}$$

is the speech distortion due to the filtering matrix and

$$\mathbf{e}_{\mathrm{r}}(m) = \mathbf{H}(m)\mathbf{v}(m) \tag{6.21}$$

represents the residual noise.

Having defined the error signal, we can now write the fullband MSE criterion:

$$\begin{aligned} J[\mathbf{H}(m)] &= \mathrm{tr}\left\{E\left[\mathbf{e}(m)\mathbf{e}^H(m)\right]\right\} \\ &= \mathrm{tr}\left[\mathbf{\Phi}_{\mathbf{x}}(m)\right] - \mathrm{tr}\left[\mathbf{H}(m)\mathbf{\Phi}_{\mathbf{y}}(m)\mathbf{H}^H(m)\right] \\ &\quad - \mathrm{tr}\left[\mathbf{H}(m)\mathbf{\Phi}_{\mathbf{x}}(m)\right] - \mathrm{tr}\left[\mathbf{\Phi}_{\mathbf{x}}(m)\mathbf{H}^H(m)\right]. \end{aligned} \tag{6.22}$$

Using the fact that $E\left[\mathbf{e}_{\mathrm{d}}(m)\mathbf{e}_{\mathrm{r}}^H(m)\right] = \mathbf{0}_{K\times K}$, $J[\mathbf{H}(m)]$ can be expressed as the sum of two other fullband MSEs, i.e.,

$$\begin{aligned} J[\mathbf{H}(m)] &= \mathrm{tr}\left\{E\left[\mathbf{e}_{\mathrm{d}}(m)\mathbf{e}_{\mathrm{d}}^H(m)\right]\right\} + \mathrm{tr}\left\{E\left[\mathbf{e}_{\mathrm{r}}(m)\mathbf{e}_{\mathrm{r}}^H(m)\right]\right\} \\ &= J_{\mathrm{d}}[\mathbf{H}(m)] + J_{\mathrm{r}}[\mathbf{H}(m)]. \end{aligned} \tag{6.23}$$

Two particular filtering matrices are of great importance: $\mathbf{H}(m) = \mathbf{I}_K$ and $\mathbf{H}(m) = \mathbf{0}_{K\times K}$. With the first one (identity filtering matrix), we have neither noise reduction nor speech distortion and with the second one (zero filtering matrix), we have maximum noise reduction and maximum speech distortion (i.e., the desired speech signal is completely nulled out). For both filtering matrices, however, it can be verified that the output SNR is equal to the input SNR. For these two particular filtering matrices, the MSEs are

$$J[\mathbf{I}_K(m)] = J_{\mathrm{r}}[\mathbf{I}_K(m)] = \mathrm{tr}\left[\mathbf{\Phi}_{\mathbf{v}}(m)\right], \tag{6.24}$$

$$J\left[\mathbf{0}_{K\times K}(m)\right] = J_{\mathrm{d}}\left[\mathbf{0}_{K\times K}(m)\right] = \mathrm{tr}\left[\mathbf{\Phi}_{\mathbf{x}}(m)\right]. \tag{6.25}$$

As a result,

$$\mathrm{iSNR}(m) = \frac{J\left[\mathbf{0}_{K\times K}(m)\right]}{J[\mathbf{I}_K(m)]}. \tag{6.26}$$

We define the fullband NMSE with respect to $J[\mathbf{I}_K(m)]$ as

$$\begin{aligned} \widetilde{J}[\mathbf{H}(m)] &= \frac{J[\mathbf{H}(m)]}{J[\mathbf{I}_K(m)]} \\ &= \mathrm{iSNR}(m)\cdot\upsilon_{\mathrm{sd}}[\mathbf{H}(m)] + \frac{1}{\xi_{\mathrm{nr}}[\mathbf{H}(m)]} \\ &= \mathrm{iSNR}(m)\left\{\upsilon_{\mathrm{sd}}[\mathbf{H}(m)] + \frac{1}{\mathrm{oSNR}[\mathbf{H}(m)]\cdot\xi_{\mathrm{sr}}[\mathbf{H}(m)]}\right\}, \end{aligned} \tag{6.27}$$

where

$$\upsilon_{\mathrm{sd}}[\mathbf{H}(m)] = \frac{J_{\mathrm{d}}[\mathbf{H}(m)]}{J_{\mathrm{d}}\left[\mathbf{0}_{K\times K}(m)\right]}, \tag{6.28}$$

$$\text{iSNR}(m) \cdot \upsilon_{\text{sd}}\left[\mathbf{H}(m)\right] = \frac{J_{\text{d}}\left[\mathbf{H}(m)\right]}{J_{\text{r}}\left[\mathbf{I}_K(m)\right]}, \tag{6.29}$$

$$\xi_{\text{nr}}\left[\mathbf{H}(m)\right] = \frac{J_{\text{r}}\left[\mathbf{I}_K(m)\right]}{J_{\text{r}}\left[\mathbf{H}(m)\right]}, \tag{6.30}$$

$$\text{oSNR}\left[\mathbf{H}(m)\right] \cdot \xi_{\text{sr}}\left[\mathbf{H}(m)\right] = \frac{J_{\text{d}}\left[\mathbf{0}_{K\times K}(m)\right]}{J_{\text{r}}\left[\mathbf{H}(m)\right]}. \tag{6.31}$$

This shows how this fullband NMSE and the different fullband MSEs are related to the performance measures.

We define the fullband NMSE with respect to $J\left[\mathbf{0}_{K\times K}(m)\right]$ as

$$\begin{aligned} \overline{J}\left[\mathbf{H}(m)\right] &= \frac{J\left[\mathbf{H}(m)\right]}{J\left[\mathbf{0}_{K\times K}(m)\right]} \\ &= \upsilon_{\text{sd}}\left[\mathbf{H}(m)\right] + \frac{1}{\text{oSNR}\ \left[\mathbf{H}(m)\right] \cdot \xi_{\text{sr}}\left[\mathbf{H}(m)\right]} \end{aligned} \tag{6.32}$$

and, obviously,

$$\widetilde{J}\left[\mathbf{H}(m)\right] = \text{iSNR}(m) \cdot \overline{J}\left[\mathbf{H}(m)\right]. \tag{6.33}$$

We are only interested in filtering matrices for which

$$J_{\text{d}}\left[\mathbf{I}_K(m)\right] \le J_{\text{d}}\left[\mathbf{H}(m)\right] < J_{\text{d}}\left[\mathbf{0}_{K\times K}(m)\right], \tag{6.34}$$

$$J_{\text{r}}\left[\mathbf{0}_{K\times K}(m)\right] < J_{\text{r}}\left[\mathbf{H}(m)\right] < J_{\text{r}}\left[\mathbf{I}_K(m)\right]. \tag{6.35}$$

From the two previous expressions, we deduce that

$$0 \le \upsilon_{\text{sd}}\left[\mathbf{H}(m)\right] < 1, \tag{6.36}$$

$$1 < \xi_{\text{nr}}\left[\mathbf{H}(m)\right] < \infty. \tag{6.37}$$

The optimal filtering matrices are obtained by minimizing $J\left[\mathbf{H}(m)\right]$ or minimizing $J_{\text{r}}\left[\mathbf{H}(m)\right]$ or $J_{\text{d}}\left[\mathbf{H}(m)\right]$ subject to some constraint.

6.3 Optimal Filtering Matrices

In this section, we are discussing three important filtering matrices.

6.3.1 Maximum SNR

Our first optimal filtering matrix is not derived from the fullband MSE criterion but from the fullband output SNR defined in (6.12) that we can rewrite as

$$\mathrm{oSNR}\left[\mathbf{H}(m)\right] = \frac{\sum_{k=0}^{K-1} \mathbf{h}_k^H(m)\mathbf{\Phi}_{\mathbf{x}}(m)\mathbf{h}_k(m)}{\sum_{k=0}^{K-1} \mathbf{h}_k^H(m)\mathbf{\Phi}_{\mathbf{v}}(m)\mathbf{h}_k(m)}. \tag{6.38}$$

It is then natural to try to maximize this SNR with respect to $\mathbf{H}$. Let us first give the following lemma.

Lemma 6.1 *We have*

$$\mathrm{oSNR}\left[\mathbf{H}(m)\right] \le \max_k \frac{\mathbf{h}_k^H(m)\mathbf{\Phi}_{\mathbf{x}}(m)\mathbf{h}_k(m)}{\mathbf{h}_k^H(m)\mathbf{\Phi}_{\mathbf{v}}(m)\mathbf{h}_k(m)} = \chi. \tag{6.39}$$

Proof Let us define the positive reals $a_k = \mathbf{h}_k^H(m)\mathbf{\Phi}_{\mathbf{x}}(m)\mathbf{h}_k(m)$ and $b_k = \mathbf{h}_k^H(m)$ $\mathbf{\Phi}_{\mathbf{v}}(m)\mathbf{h}_k(m)$. We have

$$\frac{\sum_{k=0}^{K-1} a_k}{\sum_{k=0}^{K-1} b_k} = \sum_{k=0}^{K-1}\left(\frac{a_k}{b_k} \cdot \frac{b_k}{\sum_{i=0}^{K-1} b_i}\right). \tag{6.40}$$

Now, define the following two vectors:

$$\mathbf{u} = \left[\frac{a_0}{b_0}\ \frac{a_1}{b_1} \cdots \frac{a_{K-1}}{b_{K-1}}\right]^T, \tag{6.41}$$

$$\mathbf{u}' = \left[\frac{b_0}{\sum_{i=0}^{K-1} b_i}\ \frac{b_1}{\sum_{i=0}^{K-1} b_i} \cdots \frac{b_{K-1}}{\sum_{i=0}^{K-1} b_i}\right]^T. \tag{6.42}$$

Using the Holder's inequality, we see that

$$\begin{aligned}\frac{\sum_{k=0}^{K-1} a_k}{\sum_{k=0}^{K-1} b_k} &= \mathbf{u}^T\mathbf{u}' \\ &\le \left\|\mathbf{u}\right\|_\infty \left\|\mathbf{u}'\right\|_1 = \max_k \frac{a_k}{b_k},\end{aligned} \tag{6.43}$$

which ends the proof.

Theorem 6.1 *The maximum SNR filtering matrix is given by*

$$\mathbf{H}_{\max}(m) = \begin{bmatrix} \alpha_0(m)\mathbf{h}_{\max}^H(m) \\ \alpha_1(m)\mathbf{h}_{\max}^H(m) \\ \vdots \\ \alpha_{K-1}(m)\mathbf{h}_{\max}^H(m) \end{bmatrix}, \tag{6.44}$$

where $\alpha_k(m)$, $k = 0, 1, \ldots, K-1$ *are arbitrary numbers with at least one of them different from 0. The corresponding output SNR is*

$$\text{oSNR}\left[\mathbf{H}_{\max}(m)\right] = \lambda_{\max}(m), \tag{6.45}$$

where $\lambda_{\max}(m)$ *is the maximum eigenvalue of the matrix* $\boldsymbol{\Phi}_{\mathbf{v}}^{-1}(m)\boldsymbol{\Phi}_{\mathbf{x}}(m)$ *and its corresponding eigenvector is* $\mathbf{h}_{\max}(m)$.

Proof From Lemma 6.1, we know that the fullband output SNR is upper bounded by χ whose maximum value is clearly $\lambda_{\max}(m)$. On the other hand, it can be checked from (6.38) that $\text{oSNR}\left[\mathbf{H}_{\max}(m)\right] = \lambda_{\max}(m)$. Since this output SNR is maximal, $\mathbf{H}_{\max}(m)$ is indeed the maximum SNR filtering matrix.

Property 6.1 *The fullband output SNR with the maximum SNR filtering matrix is always greater than or equal to the fullband input SNR, i.e.,* $\text{oSNR}\left[\mathbf{H}_{\max}(m)\right] \geq \text{iSNR}(m)$.

It is interesting to see that we have these bounds:

$$0 \leq \text{oSNR}\left[\mathbf{H}(m)\right] \leq \lambda_{\max}(m), \forall \mathbf{H}(m), \tag{6.46}$$

but, obviously, we are only interested in filtering matrices that can improve the output SNR, i.e., $\text{oSNR}\left[\mathbf{H}(m)\right] \geq \text{iSNR}(m)$.

If the spectral correlation of the signals can be neglected [meaning that $\boldsymbol{\Phi}_{\mathbf{x}}(m)$ and $\boldsymbol{\Phi}_{\mathbf{v}}(m)$ are diagonal matrices], then $\mathbf{H}_{\max}(m)$ is a diagonal matrix with only one nonnull element. This maximum SNR filter is obviously equivalent to the one derived in Chap. 2.

6.3.2 Wiener

If we differentiate the fullband MSE criterion, $J\left[\mathbf{H}(m)\right]$, with respect to $\mathbf{H}(m)$ and equate the result to zero, we find the Wiener filtering matrix

$$\begin{aligned}\mathbf{H}_{\mathrm{W}}(m) &= \boldsymbol{\Phi}_{\mathbf{x}}(m)\boldsymbol{\Phi}_{\mathbf{y}}^{-1}(m) \\ &= \mathbf{I}_K - \boldsymbol{\Phi}_{\mathbf{v}}(m)\boldsymbol{\Phi}_{\mathbf{y}}^{-1}(m),\end{aligned} \tag{6.47}$$

which is identical to the one derived in [4]. This matrix depends only on the second-order statistics of the noise and observation signals.

If the spectral correlation of the signals can be neglected, then $\mathbf{H}_{\mathrm{W}}(m)$ is a diagonal matrix whose components are the Wiener gains derived in Chap. 2.

Property 6.2 *The fullband output SNR with the Wiener filtering matrix is always greater than or equal to the fullband input SNR, i.e.,* $\text{oSNR}\left[\mathbf{H}_{\mathrm{W}}(m)\right] \geq \text{iSNR}(m)$.

6.3.3 Tradeoff

In the tradeoff approach, we minimize the speech distortion index with the constraint that the noise reduction factor is equal to a positive value that is greater than 1. Mathematically, this is equivalent to

$$\min_{\mathbf{H}(m)} J_{\mathrm{d}}[\mathbf{H}(m)] \quad \text{subject to} \quad J_{\mathrm{r}}[\mathbf{H}(m)] = \beta J_{\mathrm{r}}[\mathbf{I}_K(m)], \tag{6.48}$$

where $0 < \beta < 1$ to insure that we get some noise reduction. By using a Lagrange multiplier, $\mu \geq 0$, to adjoin the constraint to the cost function and assuming that the matrix $\mathbf{\Phi}_{\mathbf{x}}(m) + \mu\mathbf{\Phi}_{\mathbf{v}}(m)$ is invertible, we easily deduce the tradeoff filtering matrix

$$\mathbf{H}_{\mathrm{T},\mu}(m) = \mathbf{\Phi}_{\mathbf{x}}(m)\left[\mathbf{\Phi}_{\mathbf{x}}(m) + \mu\mathbf{\Phi}_{\mathbf{v}}(m)\right]^{-1}. \tag{6.49}$$

For $\mu = 1$, we get the Wiener filtering matrix. For $\mu = 0$, we see that $\mathbf{H}_{\mathrm{T},0}(m) = \mathbf{I}_K$. For μ greater or smaller than 1, we obtain a filtering matrix that reduces more or less noise than the Wiener filtering matrix.

Property 6.3 *The fullband output SNR with the tradeoff filtering matrix is always greater than or equal to the fullband input SNR, i.e.,* $\mathrm{oSNR}\left[\mathbf{H}_{\mathrm{T},\mu}(m)\right] \geq \mathrm{iSNR}(m)$, $\forall\mu \geq 0$.

It can also be shown that for $\mu \geq 1$,

$$\mathrm{iSNR}(m) \leq \mathrm{oSNR}\left[\mathbf{H}_{\mathrm{W}}(m)\right] \leq \mathrm{oSNR}\left[\mathbf{H}_{\mathrm{T},\mu}(m)\right] \leq \mathrm{oSNR}\left[\mathbf{H}_{\max}(m)\right] \tag{6.50}$$

and for $0 \leq \mu \leq 1$,

$$\mathrm{iSNR}(m) \leq \mathrm{oSNR}\left[\mathbf{H}_{\mathrm{T},\mu}(m)\right] \leq \mathrm{oSNR}\left[\mathbf{H}_{\mathrm{W}}(m)\right] \leq \mathrm{oSNR}\left[\mathbf{H}_{\max}(m)\right]. \tag{6.51}$$

References

1. N.L. Gerr, J.C. Allen, The generalized spectrum and spectral coherence of a harmonizable time series. Digit. Signal Process. **4**, 222–238 (1994)
2. A. Napolitano, Uncertainty in measurements on spectrally correlated stochastic processes. IEEE Trans. Signal Process. **49**, 2172–2191 (2003)
3. J. Benesty, J. Chen, Y. Huang, I. Cohen, *Noise Reduction in Speech Processing* (Springer, Berlin, 2009)
4. C. Li, S.V. Andersen, A block-based linear MMSE noise reduction with a high temporal resolution modeling of the speech excitation. EURASIP J. Appl. Signal Process. **2005**(18), 2965–2978 (2005)

Chapter 7
Summary and Perspectives

Speech enhancement fulfills a key role in a variety of applications such as, mobile phones, hearing aids, video-conferencing systems, and human-machine interfaces. By processing the observed noisy microphone signal(s), it is possible to improve the quality and in some cases also the intelligibility of the desired speech. Even after several decades of research in this area, the quest for robust and efficient speech enhancement algorithms continues to exist. Most single microphone solutions suffer from an inherent tradeoff between noise reduction and speech distortion and although they are able to increase speech quality, improvements in terms of intelligibility are rarely reported. Compared to single microphone (single-channel) solutions, multi-microphone (multichannel) solutions are able to reduce significantly more noise without distorting the desired speech. Due to the increasing demand for high-quality transparent audio communication systems and seamless human-machine interaction, it is not surprising that multi-microphone solutions are becoming more popular.

In this work, we focused on optimal linear gain functions and filters for single-channel and multichannel speech enhancement. Throughout this study, we worked in the frequency domain via the short-time Fourier transform (STFT). Major advantages of working in this domain are that it allows producing low-latency speech enhancement algorithms and that it allows exploiting short- and long-term statistics of the desired and noise signals. The theory in this work was presented in a unified manner that enables a systematic development of performance measures, gain functions, and filters.

In Chap. 2, we focused on gain functions that operate on the noisy STFT signal. A single-channel signal model and performance measures were described. The measures are not only extremely useful to evaluate the filter's subband and full-band performance in terms of the signal-to-noise ratio (SNR), noise reduction, and speech distortion, but also to develop different gain functions. From these measures, it has also become apparent that the subband SNR cannot be improved by applying a gain to each time and frequency instance and that there always exists a tradeoff between noise reduction and speech distortion. Among the developed gain functions, the tradeoff gain function is of great interest as it facilitates the tradeoff between

J. Benesty et al., *Speech Enhancement in the STFT Domain*,
SpringerBriefs in Electrical and Computer Engineering,
DOI: 10.1007/978-3-642-23250-3_7,

the amount of noise reduction and speech distortion. This particular gain function also provides a starting point for the development of novel perceptually motivated noise reduction functions that take properties of the human auditory system such as temporal masking into account.

In Chap. 3, we further generalized the signal model in such a way that it becomes possible to apply frequency dependent filters to the noisy STFT signals. This generalization makes it possible to exploit the correlation between successive STFT frames (i.e., interframe correlation). Aside form the derived conventional filters such as the Wiener filter that always introduces some speech distortion, it was possible to derive a minimum variance distortionless response (MVDR) filter that is able to reduce noise without distorting the desired speech signal. From this perspective we can also exploit the interframe correlation of the noise signals in the STFT domain. By decomposing the STFT noise signals into two orthogonal signal components, we are able to derive a linearly constraint minimum variance (LCMV) filter in the single-channel case that is able to completely cancel the noise that is correlated across the STFT frames at a specific frequency. In practice, this filter might be of interest when dealing with noise that is periodic within a subband. In contrast to the gain functions described in Chap. 2, the filters described in this chapter may in the future prove to be useful to improve both the speech quality and speech intelligibility.

When multiple microphone signals are available, we can exploit the spatial diversity in addition to the temporal/spectral diversity of the desired and noise signals. In Chap. 4, we focused on optimal gain functions that operate on the STFT of multiple microphone signals. The introduced signal model is general and can be used to describe the microphone signals in free-field and reverberant-field. Here our objective was to estimate the desired signal as received at one of the microphones. Consequently, we focused entirely on noise reduction and did not attempt to dereverberate the observed desired signal. From this perspective, it was possible to derive gain functions that only depend on the spatial correlation matrix of observed and noise STFT signals. The advantage of this approach is multiple. First, we do not require any prior information of the location of the desired source with respect to the microphone array. Secondly, we do not require information of the geometry of the microphone array. Finally, we do not require the microphone array to be calibrated. From a practical point of view, the deduced spatial prediction gain function (as derived in Sect. 4.4) is of great interest. This function depends only on the acoustic transfer function ratios related to the desired source and the spatial correlation matrix of the noise and introduces no speech distortion. Another interesting function is the LCMV gain function. Especially nonstationary noise sources provide a major challenge for speech enhancement systems because it is difficult to track the short-term temporal statistics of such noise sources. When the locations of such noise sources are fixed or slowly varying across time, we can rely more heavily on the spatial characteristics of the noise sources to cancel the noise. The deduced LCMV function described in this chapter provides a viable solution in such a scenario and does not require estimates of the acoustic transfer function ratios for each individual noise source but requires a single vector that decomposes the noise into two noise components, viz., one that is coherent across the microphones and one that is incoherent across the microphones.

In Chap. 5, we presented a new perspective on the problem of multichannel noise reduction in the STFT domain by exploiting spatio-temporal information of the desired and noise signals. It was shown that for a specific frequency the desired signal received by the first microphone at a particular time instance can be related to (a) the desired signal received by the first microphone at different time instances and (b) the desired signal received by other microphones at any time instance. The relation is described using a spatio-temporal steering vector that can be expressed in terms of a Kronecker product of the interframe correlation of the desired source (as introduced in Chap. 3) and the classical steering vector (as described and used in Chap. 4). Based on this signal model, we derived frequently used performance measures and a variety of frequency-dependent optimal linear filters. In practice, a reasonable input SNR is required to accurately estimate the interframe correlation of the desired signal. In case this information can be obtained, it is expected that we can further increase the noise reduction performance compared to more traditional gain functions as presented in Chap. 4 without increasing the speech distortion.

In the Chaps. 2–5, we worked under the assumption that the STFT signals can be processed independently for each frequency. Although this is assumption is commonly made, it is only valid when the observed time-domain signal is wide-sense stationary and invalid when dealing with nonstationary signals such as speech. One way to measure the correlation between different frequency bins is by using the bifrequency spectrum. In Chap. 6, we introduced the bifrequency spectrum in speech enhancement. We have introduced a signal model and corresponding performance measures to evaluate gain functions that are based on the bifrequency spectrum. Finally, a Wiener filter and a tradeoff filter were derived. Although the formulation exploits the correlation between all frequencies, we anticipate that only the correlation between a few neighboring frequencies can be exploited in practice as shown in Chap. 1. Although the focus was on single-channel speech enhancement, this idea can be extended to the multichannel case.

The unified framework presented in this study provides a starting point for further research in this area. Although the time-frequency analysis in this work was performed using the STFT, it should be noted that the presented theory is general. Therefore any other, possibly perceptually motivated transformation can be used. In the case of microphone array processing, we have to take the geometry of the microphone array as well as the spectral properties of the desired and noise signals into account when selecting the transformation.

Depending on the gain function or filter used, estimates of steering vectors, interframe correlation vectors, variances and correlation matrices are required. Estimation of the spectral variance of the noise (single-channel scenario) and estimation of the noise correlation matrix (multichannel scenario) provide other challenging research topics. In the last decade, significant advances have been made with regards to the estimation of the spectral variance of the noise, see for examples [1–3]. Some of these estimators are able to continuously track the spectral variance of the noise even when the desired speaker is active. Only recently, researchers have started to develop robust solutions for estimating the noise correlation matrix [4, 5]. Another research question is related to the robustness of gain functions and filters to estimation errors.

The speech enhancement performance in terms of for example speech distortion and noise reduction will depend on the accuracy of the obtained estimates. While several of the developed gain functions and filters are theoretically equivalent, their performance will differ in practice because of estimation errors. Therefore, it would be of interest to investigate systematically how their performance is influenced by estimation errors.

In the last decade we have seen an increased commercial interest in multi-microphone solutions. To date, most mobile phones are equipped with two microphones and hearing-aid devices are commonly equipped with two or more microphones. In 2009, consumer electronic producers have announced Skype-enabled televisions that are being equipped with a video camera and a small microphone array. Other manufactures are offering devices that can be connected to your television and enable video-conferencing. With an increasing demand for high-quality transparent audio communication, these applications provide major acoustic signal processing challenges. Speech enhancement solutions as described in this work, which require no calibration and do not heavily rely on the geometry of the microphone array, provide major advantages for these and other applications. A further development goes in the direction of using distributed microphones and distributed microphone arrays. Although these systems provide new practical and theoretical challenges, the optimal gain functions and filters described in this study are easily applicable.

References

1. R. Martin, Noise power spectral density estimation based on optimal smoothing and minimum statistics. IEEE Trans. Speech Audio Process. **9**, 504–512 (2001)
2. I. Cohen, Noise spectrum estimation in adverse environments: improved minima controlled recursive averaging. IEEE Trans. Speech Audio Process. **11**, 466–475 (2003)
3. R.C. Hendriks, R. Heusdens, J. Jensen, MMSE based noise PSD tracking with low complexity, in *Proceedings IEEE ICASSP*, (2010), pp. 4266–4269
4. M. Souden, J. Chen, J. Benesty, S. Affes, Gaussian model-based multichannel speech presence probability. IEEE Trans. Audio Speech Lang. Process. **18**, 1072–1077 (2010)
5. M. Souden, J. Chen, J. Benesty, S. Affes, An integrated solution for online multichannel noise tracking and reduction. IEEE Trans. Audio Speech Lang. Process. (2011)

Index

J. Benesty et al., *Speech Enhancement in the STFT Domain*,
SpringerBriefs in Electrical and Computer Engineering,
DOI: 10.1007/978-3-642-23250-3,

S

T

W